The Cambridge Manuals of Science and
Literature

THE EARTH
ITS SHAPE, SIZE, WEIGHT AND SPIN

THE EARTH
ITS SHAPE, SIZE, WEIGHT AND SPIN

BY

J. H. POYNTING,
Sc.D., F.R.S.

Cambridge :
at the University Press
1922

CAMBRIDGE UNIVERSITY PRESS
Cambridge, New York, Melbourne, Madrid, Cape Town,
Singapore, São Paulo, Delhi, Tokyo, Mexico City

Cambridge University Press
The Edinburgh Building, Cambridge CB2 8RU, UK

Published in the United States of America by Cambridge University Press, New York

www.cambridge.org
Information on this title: www.cambridge.org/9781107606043

First Edition 1913
Reprinted 1922
First paperback edition 2011

A catalogue record for this publication is available from the British Library

ISBN 978-1-107-60604-3 Paperback

*With the exception of the coat of arms
at the foot, the design on the title page is a
reproduction of one used by the earliest known
Cambridge printer, John Siberch, 1521*

PREFACE

THE aim of this book is to explain in a general way, without mathematical detail, how the shape and size of the Earth have been determined, how its mass has been measured, and how we know that it rotates, and so uniformly that it is a nearly perfect time-keeper. Some account is given of the tidal action which must gradually be reducing the spin, a subject of which our knowledge is chiefly due to the researches of Sir George Darwin.

Readers who wish to study further the matters dealt with here, will find more detailed treatment under various headings in the eleventh edition of the *Encyclopaedia Britannica*. A bibliography of each subject is there given.

On the Tides, Sir George Darwin's general account should be read in his work on *The Tides and Kindred Phenomena of the Solar System*.

J. H. P.

November, 1912.

CONTENTS

CHAPTER I

THE SHAPE AND SIZE OF THE EARTH

IF we stand on a hill top on a clear day, and look over the lowlands stretching away from below, there is nothing in what we see to suggest that we are on the surface of a globe. There is no appearance that the surface bends down from us as it recedes. Rather does it seem as if the Earth slopes up towards the horizon and as if the hill rises up in the middle of a shallow cup.

When men first began to think about such observations as this, and to consider the shape of the Earth, there was no obvious suggestion that they were on a globe, and, naturally perhaps, the first idea was that the Earth is a flat plain on which the mountains are creases, a flat 'firmament in the midst of the waters.'

Gradually, however, observations accumulated which could not be reconciled with the flatness of the Earth. A traveller, journeying from a mountain range, found on looking back that the mountains not only grew smaller and smaller but that they sank and at last dropped down altogether out of sight.

When men began to go down to the sea in ships and ventured far out on the waters, the new land to which they sailed appeared first as one little peak, then as a range, and at last the whole land stood above the water. These observations were difficult to reconcile with the idea of a flat Earth, but easy to explain if it was round.

The doctrine of the roundness of the Earth, then, gradually replaced the doctrine of its flatness. But there was a long fight of nearly 2000 years between the doctrines. When Columbus at the end of the 15th century proposed to reach India by sailing to the west instead of to the east, arguing that as the Earth was round, the other side might be reached either way, his opponents, holding that the Earth was flat, regarded him as a fool and a heretic. It was urged that if the Earth were round, men on the opposite side would be walking with their heels upwards, that the trees would be growing with their branches downwards, and that it would rain, hail and snow upwards. All this appeared to them absurd, for they did not realise that the tendency to fall is a tendency to fall towards the centre of the Earth. They thought of 'falling' as a motion in the same direction everywhere, and anything loose on the other side of the Earth, if that other side could be conceived as existing, should fall away from the surface. They argued that, in order to travel from that other side to this, a ship would have to climb up the sea as if it were climbing up a mountain slope, and that no wind would suffice to drive it up. It was even urged that the roundness of the Earth was inconsistent with the resurrection of the body. For the dead on the other side of the globe would rise on this side with their heels uppermost.

Columbus fought the last fight against a flat Earth, and won. He sailed to the west and found, not indeed the India which he had hoped for, but the West Indies. Soon after, the journey round the world was made, and the Earth was henceforth a globe for all who could study the evidence. Let us consider this evidence in its most conclusive form.

If we watch the stars, by night, at a place in this part of the world, we see that one star, the pole star, does not noticeably change its position, and that all the other stars circle round it. When we make careful measurements we find that the pole star is not quite fixed but goes in a small circle round a centre, which we may conveniently call the sky pole, and it is this sky pole round which all the other stars circle.

If we travel northwards, the stars still circle round the same pole, but the pole itself rises higher in the sky. The fundamental fact is, that for the same distance of travel due north, the pole rises the same number of degrees, wherever our starting-point may be.

If we are at sea where the horizon is definite we may measure the height in degrees above that horizon. If we are on land where the horizon is conditioned by the elevation of the land and is therefore not a definite line, we may measure the distance in degrees from the zenith, the point directly overhead ; and the zenith distance is 90° *minus* the distance from the true horizon, where, by the true horizon, we mean that which the sea-line would give if we had sea in place of land. Thus at a point near Nottingham the sky pole is 53° above the true horizon or 37° below the zenith. If we travel due north 69 miles, to a point near York, the sky pole is 54° above the true horizon or 36° from the zenith. Or if we go across to Ireland, at Cavan it stands 54° above the true horizon, while 69 miles due north at Londonderry it stands 55° above the horizon. Or taking a longer distance, at Coventry it stands about 52½° above the horizon, while at Sandwick in the Shetlands 552 miles due north it stands about 60½° above the horizon, having risen 8° for a travel 8 × 69 miles northward. Everywhere the rise is very nearly at the same rate of 1° for 69 miles' travel northwards, not exactly the same, as we shall see later, for the distance increases slightly as we go from the equator towards the pole, but the increase is very slight.

Postponing for the present the description of the way in which pole height and distances are measured we may see at once that the relation we have stated is quite inconsistent with a flat Earth. Let us take the last case of Coventry and Sandwick and for simplicity of statement let us think of the pole star as actually at the sky pole. It need not affect the conclusion, for at one point in its circle the pole star will be at the same height above the horizon as the sky pole, and we may choose that point for consideration.

Let C (fig. 1) represent Coventry on a flat Earth, and S Sandwick 552 miles due north. At C make the angle $PCN = 52\frac{1}{2}°$ and at S make the angle $PSN = 60\frac{1}{2}°$, the two lines CP and SP meeting in P the pole star. It is easy to calculate by trigonometry, or to find by direct measurement of a carefully

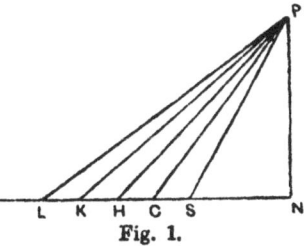

Fig. 1.

drawn figure in which CS represents 552, that to the same scale PN is about 2740 and CP is about 3450. That is, a flat Earth requires that the pole star is less than 3500 miles from Coventry, an utterly absurd result as we shall presently see. But passing this by, let us mark points *H, K, L*, distant 552 miles, 2 × 552 miles, and 3 × 552 miles respectively from *C*. On a flat Earth the angles *PCN, PHN, PKN, PLN* do not decrease successively by equal amounts. Careful measurements on a large figure suffice to show this. On the real Earth they decrease successively by 8°, so that the Earth cannot possibly be flat.

Another set of measurements would show equally well that the Earth is not flat, and they are worthy of description inasmuch as they give us conclusive evidence of the true form.

If the Earth were flat and the pole star were vertically above a point *O* (fig. 2 (*a*)) on the flat sur-

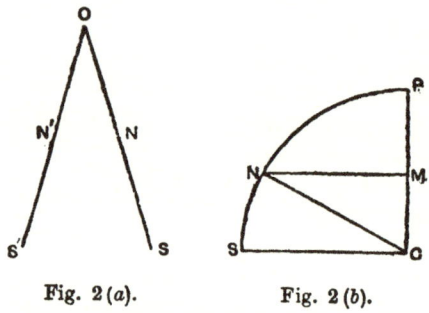

Fig. 2 (*a*).   Fig. 2 (*b*).

face, two lines of travel *SN, S'N'*, each towards due north, would be straight lines and would meet at

O. The distances between the two lines at *SS'* or *NN'* would be proportional to the distance along either from *O.* Two travellers along these lines would approach each other by equal amounts for equal distances travelled northward. But the law of approach is quite different. If we start from two points on the equator, the line on the Earth's surface for which the sky pole is on the horizon, and travel due north from each, that is, always towards the point on the horizon immediately under the sky pole, the distance between the two lines of travel, measured along a line of equal pole height, is proportional to the cosine of the angle through which the pole has risen in the sky. Or if we draw a quarter circle (fig. 2 (*b*)), and represent the distance travelled along each line by the length *SN* along this circle, the angle *SCN* being the rise of the sky pole, then the distance between the lines, if measured along a line of equal pole height, will be proportional to *NM.*

I have thought it worth while, even though the doctrine of a flat Earth has long been abandoned, to examine carefully the evidence which led to its abandonment. For that examination enables us to see that our ancestors were not so wrongheaded in holding the doctrine as, at first thought, they might seem to have been. They accepted the most obvious account of appearances, a flat Earth. They observed that everything tended to fall straight down to the Earth ; everywhere, as far as they could tell, in the same direction; or down-ness was universal. Having once taken this view, it was a real difficulty, rightly felt, that bodies could remain on the surface at the Antipodes. They should fall down into space there just as they fall down to the surface here. They had no measurements contradicting their view, nor had they means to make the measurements had they wished to do so.

Another piece of evidence, commonplace to us, was utterly closed to them. They had no difficulty in thinking of the Sun as rising up over the edge of the Earth and illuminating the whole surface at once. It was only with the invention of trustworthy portable clocks that it could be clearly proved that sunrise, noon, and sunset take place earlier and earlier the further we go eastward, later and later the further we go westward. Every traveller across the Atlantic knows that the clocks on board have to be altered each night to make them agree even fairly with the Sun, and every follower of cricket knows that the Australians may have finished a match before we breakfast.

We shall now examine the interpretation which we are obliged to give to the two sets of measurements which we have described, viz.:

1. That the sky pole rises through equal angles

towards the zenith for equal distances travelled due northwards.

2. That two lines on the Earth's surface, each drawn due northwards, are a distance apart, if measured along a line of equal pole height, proportional to the cosine of the pole height or proportional to the length NM in fig. 2 (*b*).

It is first necessary to show that we can fix a definite direction in space, wherever we may be on the Earth's surface, by drawing a line to one of the fixed stars.

Wherever we may be on the Earth, if we see the same groups of stars those groups form the same patterns in the sky. This shows at once that they are vast distances away. For if we look at any arrangement of objects, the less does the arrangement appear to change with a change in our position the further the objects are from us. As we walk along a road, the view of a house by the roadside changes almost with every step. A wood further back alters more slowly. Still, as we move the trees do appear to change places, a nearer tree being now in front of one, now in front of another of those further back. But a range of distant hills may show just the same appearance even though we move hundreds of yards along the road. No measurements which have been made, even with the most powerful telescopes, from different parts of the Earth's surface

at the same time have ever shown the least difference in pattern of the stars in any constellation,
and we are forced to conclude that the stars are
immensely distant in comparison with any distance
we can set out on the Earth. Indeed, the pattern
only changes very minutely if we use the vastly
greater distance afforded by the motion of the Earth
round the Sun from one side of its orbit to the other.

It follows that the direction of any one fixed star
is, as nearly as we can measure, the same as seen
from all parts of the Earth, or that straight lines
drawn from all points to the star are parallel. This
will hold good if, instead of any particular star, we
take the point about which the stars in their patterns
appear to circle; that is to say, the sky pole.

Let S (fig. 3) be a point from which the sky pole
is seen along the line SP, and let N be a point due
north of S from which the pole is seen along the line
NP' parallel to SP. If SZ and NZ' are the verticals
at S and N, that is, the lines directed towards their
respective zeniths, the angle ZSP is greater than the
angle $Z'NP'$, and, as we have seen, if SN is 69 miles
it is greater by 1°. The surface therefore bends
away from a fixed direction, that of the sky pole, by
equal amounts in equal distances. This shows at
once that in going northwards we are travelling in a
circle, for that is the only curve which bends through
equal angles in equal distances. If we produce the

two verticals ZS, $Z'N$ to meet in O, O is the centre of the circle. If we produce $P'N$ to meet OS at R, the angle

$$NOS = NRS - ONR = PSZ - P'NZ'.$$

If then $SN = 69$ miles, $NOS = 1°$, and since there are 360° in the complete circle its circumference is

$$69 \times 360 = 24,840 \text{ miles,}$$

and the radius is 3950 miles, since the circumference of a circle is 6·283 × radius, very nearly. These numbers are not quite exact, since the distance 69 miles for 1° rise is not quite exact.

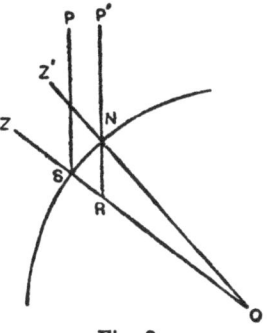

Fig. 3.

We have supposed that we are travelling northwards where the northern sky pole is visible. But if we travel southwards beyond the equator, where the southern sky pole is visible, we have the same rise of 1° per 69 miles travel, so that we move in a circle of the same size everywhere.

The surface of the Earth must therefore have a shape such that a plane drawn through the vertical at any point, and through the line to the sky pole, must cut the surface in a circle with radius about

3950 miles. There are three and only three shapes for which this could be true, a cylinder, like a round ruler, an anchor ring, and a sphere. The second set of measurements on p. 9, giving the law of approach of two lines both running due north, at once enables us to decide between the three. On a cylinder the two lines would always be the same distance apart. On an anchor ring they would approach, but more slowly than the measured rate. On a sphere alone would they approach at the measured rate. We may easily see that a sphere gives this rate. For if PP' (fig. 4) is the diameter of the sphere parallel to the direction of the sky pole, and if two planes through PP' cut the surface in the circles $PNEP'$ and $PN'E'P'$, let

$$PN = PN',$$

and let MN, MN' be perpendiculars to PP'. Let NN' be an arc of a circle with centre M. If α is the angle between the two planes measured in radians,

$$NN' = NM \cdot \alpha,$$

or NN' is proportional to the length NM. If we denote by λ the angle NOE, where E is halfway between P and P', then

$$NM = ON \cos \lambda.$$

Hence NN' is proportional to $\cos \lambda$, and λ is easily seen to be the height of the sky pole above the horizon.

Our measurements, then, lead us to the irresistible conclusion that the Earth is, at least very nearly, a round globe, with a radius about 4000 miles. If we draw circles round it passing through P and P', and running due north and south, they are lines of longitude. If we draw circles round it with their centres at different points in PP', they are lines of latitude. In fig. 4 the angle NOP is the angle between the vertical at N and the direction to the sky pole. The angle

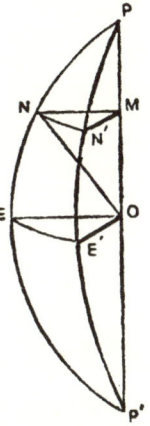

Fig. 4.

$$NOE = 90° - NOP$$

is the angle which the sky pole makes with the horizon at N, and since NOE is termed the latitude of N, the height of the sky pole above the horizon at a place is equal to the latitude of that place.

The determination of the size of the globe depends on the measurement of the angle which the sky pole makes with the horizon or with the zenith, and on the change in this angle when we travel measured distances north or south. We must now consider how we can assert that, for instance, the pole stands $52\frac{1}{2}°$ above the horizon at Coventry and $60\frac{1}{2}°$ above it at Sandwick, and how we can measure the distance between these two stations and assert that it is 552 miles.

First, as to the measurement of the pole height. We may suppose that for this purpose we use a theodolite, an instrument which is represented diagrammatically in fig. 5. *BB* is a tripod base on

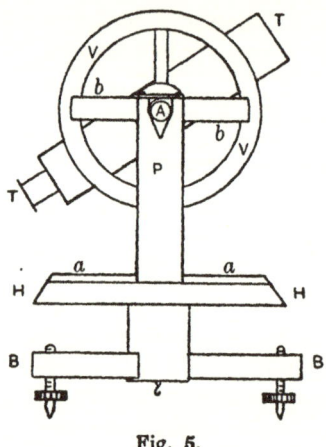

Fig. 5.

levelling screws. Only two feet of the three are shown. On the base is fixed a horizontal circular plate *HH* divided to degrees and fractions. Above this is a framework essentially consisting of two pillars, of which the front one only, *P*, is shown. This framework is mounted on an axis which fits a vertical bearing in the tripod base, and attached to it are two arms aa with verniers on them marking the position of the framework on the horizontal circle HH. At the tops of the pillars are two V bearings for a horizontal axis A, which carries a telescope TT. On this telescope is fixed a vertical circle VV divided to degrees and fractions. Two arms bb with verniers on them are fixed to the pillar P, and as the telescope revolves, and carries the circle VV with it, these arms mark the angle on the circle through which the telescope has revolved. In the telescope are two cross wires or some equivalent arrangement, so that the image of the object looked at may be brought exactly to the same point, always in the middle of the field of view. We need not enter into the modes of adjusting the two axes so as to be respectively vertical and horizontal. These will be found in any book on surveying.

Now suppose that the telescope is directed to the pole star, and that its position on the vertical circle is read. If it can be then turned round exactly into the horizontal direction and its position again be read the difference will give the height of the pole star above the horizon. But it is only at sea, by day, that we have a definite horizon to turn to, and even then allowance must be made for the fact that the line from a point any distance above sea-level slopes downwards to the horizon, owing to the curvature of the Earth. We can, however, entirely dispense with the horizon by using a horizontal reflecting surface such as is afforded by a small trough of mercury. One way of using the mercury-trough consists in placing it between the pillars of the theodolite, and pointing the telescope vertically down towards it. It is known when the telescope is exactly vertical by observing when the reflexion in the mercury of the cross wires in the eye-piece coincides with the actual cross wires, a special illuminating device, which we need not describe, being used to make the cross wires and their reflexion visible. The position of the telescope is then read on the vertical circle, and when it is turned through 90° from this position it is pointing to the true horizon.

But the pole star is not exactly at the pole of the sky. It circles round it. If, however, we measure its height when at its highest point above the horizon and its height when at its lowest point, and take the mean of these, we get the height of the centre round which it is circling.

We have taken the pole star as an example of the method of determining the pole height. Any one of a large number of stars would serve equally well, for their angular distance from the pole is accurately known from measurements which have been made at observatories, such as that at Greenwich. If then we measure the height of one of these stars when it is crossing the meridian, and therefore at its highest or lowest point in the sky, it will easily be seen that we may deduce the height of the pole.

There is a correction to be made to the observed height of a star owing to the fact that light does not come straight through the atmosphere unless it comes from the zenith, but bends down somewhat. The direction in which a star appears to be is the direction in which a ray from it enters the telescope. The star therefore is not quite so high in the sky as it appears to be. This displacement has been determined by finding what correction must be made to the observed heights of a star as it circles round the pole to make them all fall on the same circle, and tables are made giving the correction to be applied to the observed height to turn it into the true height for every position of a star.

Now as to measurement of distances on the Earth's surface. How is the size of a county, a kingdom or a continent determined? We might chain lengths as a surveyor chains a small plot of land, but the labour for any great distance would be immense and the undulations of the ground would bring in errors of very considerable amount.

Fortunately there is a method which enables us to measure distances from one point to another hundreds of miles apart with an error hardly more than a few feet. This is the 'base-line' method, and it depends upon the fact that in all triangles with the same three angles, the sides are in the same proportion to each other, so that if in any one triangle we know the length of one side which we will take as the base, and if we know the number of degrees in each of the two angles at the base, we can calculate the lengths of the other two sides of the triangle by known rules of trigonometry without further measurement.

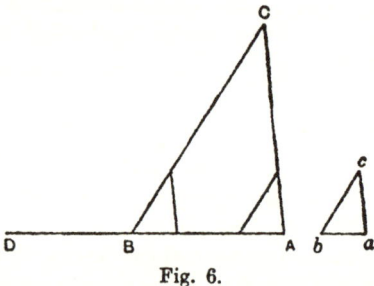

Fig. 6.

We may illustrate the principle of the method by a very simple case. Suppose that we wish to measure the distance between two points A and C (fig. 6), say on opposite sides of a river, without crossing the river. Let A be on the observer's side. The observer is to cut a triangle out of cardboard abc. He must fix the corner a at A, so that looking along ac he sees C, while on looking along ab he has a straight course which he can traverse towards D. Having marked the line AD he moves along it, always keeping the base ab of the card in the line AD until he finds that the corner b has come to a point B, such that looking along bc he sights C again. It is obvious that the triangle BAC is similar to the triangle bac, so that AC bears to AB the same ratio that ac bears to ab. We have therefore

$$AC = \frac{ac}{ab} AB.$$

If then the observer measures ac, ab and the base AB, he can at once calculate the distance AC.

We can easily see that if we make a mistake in the angle at B it leads to a much more serious error in the measurement of AC, when the base line AB is small, than when it is not very different in length from AC. Let us take the two cases represented in fig. 7, (a) with a base line comparable with AC, (b) with a very short base line, and let us suppose that we do not move quite to the right point B but go by mistake only to B', making an error in the angle equal to BCB', about the same in each case, so that BB' will not be very different in the two. The error in the value of AC will be

$$\frac{ac}{ab} BB' = \frac{AC}{AB} BB'.$$

Since $\frac{AC}{AB}$ is much greater in fig. 7 (b) than it is in fig. 7 (*a*), the error is obviously greater in the former. Hence the base should not be very small compared with the distance to be measured if accurate measurement is to be made.

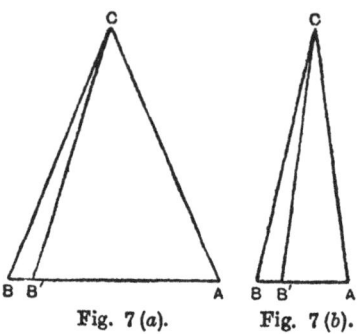

Fig. 7 (*a*).   Fig. 7 (*b*).

In applying the principle to Earth measurements, that is, to measurements such as are made in our Ordnance Survey, the first step is to choose a level surface on which a straight course may be traversed between two points *A* and *B* several miles apart. The distance *AB* constitutes the 'base-line' and it is measured as accurately as possible. Every subsequent measurement of distance depends on this first measurement. We know the size of a country the size of the Earth, the width of its orbit round the Sun, the distances of the planets and the fixed stars in terms of a base line on the surface of the Earth.

In one important respect, to be described below, the method of measuring a base-line has been changed lately. But the method is still the same in principle as the older methods, and we shall describe these, as they bring into prominence the difficulties to be overcome. If the oldest method is to be followed, two or more rods, each several yards or metres long, and usually of metal, are employed. We will suppose that we have two of these. The exact length of each rod is determined beforehand by comparison with a standard yard or metre in a

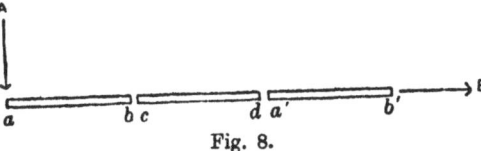

Fig. 8.

laboratory. One rod *ab* (fig. 8) is placed on supports with its end *a* at the end *A* of the base line and its length is adjusted as exactly as possible in the line *AB*. The other rod, *cd*, is then supported in the line of continuation of *ab* with a small gap between *b* and *c*, so that there shall be no risk of displacement of *ab* by contact with *cd*. The width of the gap has been measured in various ways. In a way once used a little graduated wedge was dropped into it with the narrow end downwards and the depth of its descent gave the width of the gap.

Another way is applicable to bars in which the length used is not that between the two ends of the bar but that between two marks on its upper surface. Two microscopes are fastened together with their axes a known distance apart. One of these sights the mark on ab near b, the other sights the mark on cd near c, and if the marks do not appear exactly in the middle of the fields of view allowance can be made. The gap or interval having been measured, ab is taken up and then put down on supports in the position $a'b'$ beyond and in a line with cd. The gap da' is measured and then cd is taken up and put down beyond $a'b'$, and so on, until B is reached. A and B may be marked by fine lines ruled on metal plates and the length of AB is the sum of the lengths occupied by the bars in all their positions plus the sum of the widths of the gaps.

Inasmuch as metal bars in general expand with rise of temperature, each bar or rod used in the older measurements had its length determined at some standard temperature and the change in length with any change in temperature was also measured. When a bar was being used elaborate precautions were taken to ward off inequalities of temperature in different parts of the bar and great changes of temperature from the mean, the bar being usually contained in a long double box open at the ends.

Later, Colonel Colby devised a measuring rod

for the Indian Survey which was compensated for temperature changes on a principle first used for pendulums by Ellicott.

Fig. 9 represents the skeleton of the apparatus. Two bars of different metals are used, one with considerably greater expansion than the other. For simplicity, let us suppose their expansions per degree rise to be as 3 : 2, about the ratio for brass and iron. AB is the more and CD the less expansible, and

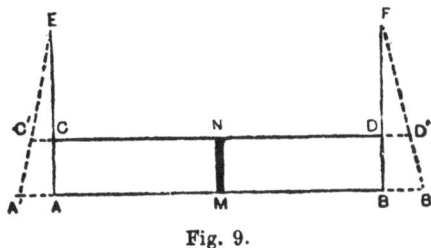

Fig. 9.

they are firmly fixed at their middle points to a cross bar MN. At their ends are rods ACE, BDF jointed or pivoted at AC, BD.

Let the apparatus at first be as indicated by the continuous lines in the figure. Then let expansion take place as indicated by the dotted lines, to $A'B'$, $C'D'$.

Since $AA' = \frac{3}{2}CC'$, $A'C'$ prolonged will cut AC in E where $AE = \frac{3}{2}CE$ or $AE = 3AC$, i.e. E is at a fixed distance along AE. So too F is at a fixed distance along *BF* and the distance *EF* remains unchanged by the expansion.

Such a compound rod requires, for successful action, that the two bars shall have the same temperatures, or at least the same temperatures at equal distances from *MN*. In practice these rods have hardly borne out the expectations formed for them at the beginning.

Now there appears to be a prospect that compensated measuring rods, like compensated pendulum rods, will be entirely superseded by wires or tapes of 'invar,' an alloy of nickel and steel discovered by M. Guillaume, which hardly changes its dimensions with ordinary changes of temperature. The wires or tapes are very much longer than the rods—as much as 100 feet—and each is stretched when in position by a definite and constant pull. Small variations of temperature from point to point are unimportant. The process of measurement becomes less cumbrous and the time required is much less[1].

Base-lines several miles in length are measured so accurately, either by the older or newer methods, that several repetitions of the measurement agree together within a fraction of an inch.

When the length of the base line *AB* (fig. 10) has

[1] *An account of the measurement of a Geodetic Base Line at Lossiemouth, in* 1909. Ordnance Survey Professional Papers, New Series, No. 1.

been determined the next step is to use it as the base of a triangle ACB, of which the vertex C is some distant but easily seen point, such, for instance, as a mark on a staff on the top of a church tower. A theodolite, the instrument represented in fig. 5, is placed at A and when the telescope sights B, the position on the horizontal circle is read. Then the telescope is moved round to sight C and the horizontal circle is read again. Thus the angle CAB is known, as it is the angle through which the telescope is turned. Then the same process is carried out at B and the angle CBA is known. Then having the length of AB and the angles at its ends we can, by the aid of trigonometry, calculate the lengths of AC and BC.

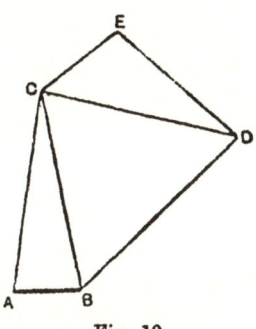

Fig. 10.

Either of these lines may be used as a base for a new triangle. For instance, BC may be used in a triangle CDB where D is perhaps a staff on a pile of stones on a hill top. The theodolite is used at B and C to measure the angles CBD and BCD. Then BD and CD can be calculated in terms of BC, and as this is known they too are known. Either of these may be used as a base for a new triangle, CD for instance, carrying us to a new point *E*. So gradually a whole country or even a whole continent may be covered with a network of triangles, and all the sides of all the triangles are found in terms of the base-line. This process is known as triangulation and, when it has been carried out, it is only a matter of trigonometrical calculation to determine the distance between any two points in the network, however far apart.

For simplicity the method has been described as if all the points lay on a flat plain. But in reality the measurements are not quite so simple as if this were the case. Thus if in the triangle *BCD*, *CB* and *CD* are not horizontal we do not measure exactly the angle *BCD*. To do that the axis of the theodolite would have to be tilted slightly so as to be perpendicular to the plane *CBD*, an adjustment which could not be made accurately even if it were desirable. But we can adjust the axis accurately in the vertical at *C*. In turning the telescope round a vertical axis from sighting *B* to sighting *D*, we really measure the angle between the vertical planes through *CB* and *CD*, not quite the same thing as the angle *BCD*. To find the latter we also have to observe on the vertical circle how much the telescope is tilted from the horizontal. There are rules which would enable us, from these measurements, to determine the angles in the triangles and the length of side. But the plan actually followed consists in projecting the straight line triangles down on to the curved surface which the ocean would give if there, that is, to the sea-level surface.

The observations enable this to be done and the network of actual triangles is replaced by a network of spherical triangles bent so as to fit the surface at sea level.

Even in getting the directions of the straight lines between the stations there is another troublesome correction to be made. A ray of light only passes straight through the air when it comes from over-head. In all other cases it is curved and therefore an object appears in a slightly different direction from that in which it would be if the air were removed. This effect of the air, the error of refraction, has been studied and can be allowed for.

The Ordnance Survey in this country began in 1784, with the measurement of a base-line on Hounslow Heath about five miles long. The original idea was to form from this base a network of triangles over the southern counties to the neighbourhood of Dover, whence it could be carried across the Channel to France. There a similar network was being formed and when the two were connected so as to form one system the difference in longitude between Greenwich and Paris, the ultimate aim, could be determined. This was soon effected, but fortunately

the work did not stop here.   The government decided to continue the triangulation over the whole of
the British Isles, and so began the great survey of
the kingdom which was only completed in 1852.   In
its course other base-lines were measured, as for
instance one on Salisbury Plain nearly seven miles
long, and one on the shores of Lough Foyle nearly
eight miles long.   Triangles were formed from the
Welsh and Scotch mountain tops to the tops of Irish
mountains, and from the north of Scotland and
Orkney to Fair Island and Foula and so on to
Shetland, and so one triangulation embraced the
whole kingdom.

As a test of the accuracy of the work a series of
triangles was selected, starting from the Lough Foyle
base and ending in a triangle of which the Salisbury
Plain base formed one side.   The length of the latter
base could then be *calculated* from the measurement
of the former and the measurements of all the angles
in the intervening triangles.   The calculation differed
from the actual measurement by less than 5 inches.
With other pairs of bases the same kind of agreement was obtained.   All the lengths calculated in all
the triangles are therefore in all probability more
accurate than 1 in 10,000.

As we have seen, the original aim was to connect
up with a continental survey.   This connection has
been repeated, and our triangulation now forms part

of a network covering all Europe. India and South
Africa have triangulations which will extend, and
at no distant date one system will no doubt spread
over the three continents of the Eastern hemisphere.
Another triangulation will cover the western continents, and distances will be known between points
separated by nearly half the Earth's circumference.
Meanwhile it suffices for the purpose of determining
the size of the globe that we are enabled to find the
exact distance between such points as Sandwick and
Coventry with a difference of pole height of 8°.

So far we have only considered the measurements
as showing that the Earth is round. It is very nearly
but not quite round. Sir
Isaac Newton showed that
if it were liquid, the spin
round its axis once in 24
hours should make it bulge
slightly at the equator and
draw in slightly at the
poles. A section through
the axis would therefore

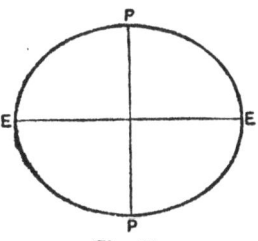

Fig. 11.

not be a circle but an oval. In fig. 11 the departure
from the circular form is enormously exaggerated,
but the exaggeration enables us to see at once that
the curve bends round more in a given distance in
the equatorial regions *EE* than in the polar regions
PP. Or if the Earth has the shape which Newton

assigned to it the vertical changes as we travel north
more rapidly in the neighbourhood of E than in the
neighbourhood of P, and the length of a degree
of latitude is less near the equator than near the
pole.

But some measurements made by Cassini early in
the 18th century appeared to show that the length
of a degree of latitude was less in the northern part
of France than in the southern part, and a school of
astronomers maintained that the earth was elongated
towards the poles—lemon-shaped instead of orange-
shaped.

It was difficult to resist the reasoning of Newton,
reasoning which would apply to a plastic solid earth,
as well as to a liquid earth, but Cassini's measurements
were against the result.   To decide the question the
French Academicians sent out two expeditions, one
to Peru in 1735 and the other to Lapland in 1736, to
determine the length of a degree of latitude in each
region.   The Peruvian expedition selected a district
at the equator near Quito, and the Lapland expedition a district near Tornea about 60° N. lat., which
was as near the pole as was convenient.   In each
case a base-line was carefully measured (in Peru two
were measured, the second one for verification), and
from it a triangulation was carried out, so that the
length of a certain line running N. and S. was determined, in Peru about 200 miles, in Lapland over 60

miles.   The change in the vertical between the ends
of these lines was measured by astronomical observations and the results were that near Quito

> one degree of latitude = 56753 toises,

and near Tornea

> one degree of latitude = 57438 toises,

a toise being about 6 feet.

Though there was some uncertainty about the
Lapland value there could not be any doubt that the
northern degree was the greater, and so it was definitely decided that the figure of the Earth was
more nearly that predicted by Newton than that
which Cassini believed to be given by his measurements.

There is an interesting story about the Peruvian
base-line near Quito.   De la Condamine, a member of
the expedition, erected two small pyramids exactly
at the ends of the base-line, so that its position
should be permanently recorded.   But soon after
his return to France he learned that the Spanish
government, probably in disapproval of the inscrip-
tions, had ordered the pyramids to be destroyed.
Subsequently orders were given for their re-erection.
Whymper in his *Travels amongst the Great Andes*
(p. 292) tells how he visited the re-erected pyramids
in 1880.   'The pyramid (of Oyambaro or Oyambarow)
which now approximately marks the southern end of
the base is about 1000 feet distant from the place

where the stone reposes [the stone on which was the offending inscription], situated in a field of maize, and is neither the original pyramid nor the one which was erected to replace it. I was informed on the spot that it was put up about thirty years earlier by a President of Ecuador, who so little appreciated the purpose for which it was originally designed that he moved it some hundreds of feet on one side, in order, he said, that *it might be better seen.*'

Subsequent measurements made in all parts of the world do not exactly fit in with any simple mathematical figure. A section through the axis of rotation is nearly but not quite an ellipse, and a section through the equator is nearly but not quite a circle. The departures from these regular figures are, however, very small—at the equator at sea-level not apparently nearly so much as a mile. We shall be making only a very minute error if we think of the section through the polar axes as an ellipse with the polar axis shorter than the equatorial axis by 1 in 293, and the surface as having the form made by the rotation of this ellipse round its shorter axis. If there were open sea at the poles, the axis from sea to sea would be, according to Col. Clarke (*Geodesy,* p. 319), 7899¼ miles, while the equatorial diameter from sea to sea is 7926½ miles, probably within ¼ mile in each case.

The Earth, then, is very round. If an exact model were made the size of a two-inch billiard ball, we should just be able to see that it was flatter at the poles, and, no doubt, in rolling it would exhibit its want of roundness. The highest mountains would be represented by elevations of $\frac{1}{800}$ inch, say by the thinnest smear of grease, the deepest oceans by the spreading of a drop into a film but $\frac{1}{700}$th inch thick.

To sum up, we find the size and shape of the Earth by measurements of lengths on its surface, starting from a base-line, and by measurements of the angles which some of the fixed stars make with the zenith when crossing the meridian. In making the astronomical measurement, it is assumed that the stars observed are so far off that lines drawn to a given star at the same instant from different parts of the Earth's surface may be regarded as parallel. This is justified by the observation that the patterns made by the constellations do not show any appreciable change when looked at, at the same time, from places as wide apart as we can have them, and by choosing stations on opposite sides of the Earth we can have them nearly 8000 miles one from the other. If further justification were needed, it would be afforded by the fact that the distances of many of the nearer fixed stars have been measured, and that these distances are so enormously great compared with the 8000 miles diameter of the Earth that the want of parallelism in lines to a star seen from opposite sides of the Earth is utterly insignificant.

We shall conclude this chapter by a short account of the way in which the distances of the nearer stars are found. It is again a base-line method, but the base is the diameter of the Earth's orbit and two stations can therefore be used 180 million miles apart. From stations so widely separated the pattern of the constellations does change slightly in some cases, the brighter and presumably nearer stars shifting slightly on the background of the fainter and presumably remoter stars as we move in the course of six months from one end to the other of the vast base-line stretching across the orbit.

Let us suppose that we have selected a star for examination, and that near it, and seen at the same time in the telescope, are faint stars which do not change their relative positions and so are presumably enormously distant. Let AB (fig. 12) be two positions of the Earth on opposite sides of the sun S, six months apart in time, 180 million miles in distance. Let C be the star to be examined and, for simplicity, we will suppose that CS is perpendicular to ASB. Of course the figure enormously exaggerates the angle ACB. It is far too minute with any actual star to be shown in a figure. If, now, we can measure[1] the angle ACB

[1] It is usual to give in tables the value of $ACS = \frac{1}{2} ACB$ and this is termed the parallax of the star.

we can at once determine the distance AC or BC, for it can be shown that

$$AC = \frac{\text{38 million million miles}}{\text{number of seconds of arc in } \overline{ACB}}.$$

Let there be a star seen near C, and in the plane of ACB, which shows no sign of change of position with regard to its fainter neighbours and let AD_1, BD_2 be lines drawn to it, presumably parallel as far as any possibility of observation goes, and let CD_3 be a third parallel.

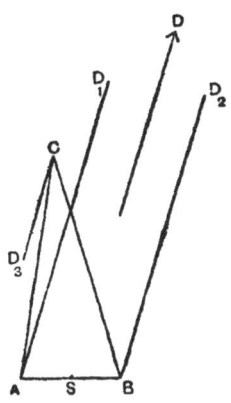

Fig. 12.

Since CD_3 is parallel to BD_2 the angles D_3CB and CBD_2 are equal to each other. And since CD_3 is parallel to AD_1 the angles D_3CA and CAD_1 are equal to each other. But

$$ACB = D_3CB - D_3CA$$
$$= CBD_2 - CAD_1.$$

In one method there is, in the eye-piece of the telescope used, measuring apparatus by which the angle CAD_1 can be measured and then, some six months later, the angle CBD_2; or photographs may be taken and the positions of the stars on these may be measured with a microscope. But even with

3—2

the nearest fixed star, α Centauri, a bright star in the southern hemisphere, ACB is only $1\frac{1}{2}$ seconds. Very minute errors in the measurement of CAD_1 and CBD_2 may produce very serious errors in the result, and it is only comparatively lately that the measurements made for the same stars by different observers have shown good agreement.

The distance of α Centauri is, from the formula given above, about 25 million million miles. Obviously the mile is an inconveniently small unit for the expression of so vast a distance, and in preference astronomers use as the unit the distance which light travels in one year, at the rate of 186,000 miles per second, or 5·8 million million miles. Thus light takes $25 \div 5\cdot8 = 4\frac{1}{3}$ years to come from the star to us. The distance is conveniently expressed as $4\frac{1}{3}$ light-years.

The distance of our own brightest star, Sirius, is about 8 light-years, while the pole star is about 63 light-years away. But here we are getting to the limit of present measurements so that it is very probable as methods improve such distances as that of the pole star will be revised. Yet, vast as is the distance which even light takes 60 years to traverse, we must regard the pole star as one of our neighbours when we compare it with other faint yet still visible members of our system.

CHAPTER II

WEIGHING THE EARTH[1]

THE Earth as a whole has no weight, if we use weight in its strict sense of earth-pull. Corresponding to each piece of the Earth here pulled down towards the centre there is another piece at the antipodes pulled up towards the centre with an equal and opposite force, and the whole globe can be divided into such neutralising pairs, leaving, of course, no outstanding pull. When, therefore, we speak of weighing the Earth we do not mean finding its weight. We mean really finding another quantity, the Mass of the Earth.

Let us first, then, try to make clear what the mass of a body is and how it is related to its weight.

If we take a pound of matter, say a piece of iron stamped as 1 lb., to different places on the surface of the Earth, we regard it as still the same pound of matter, wherever it is. Yet the earth-pull on it changes its direction, and even its amount. If we carry it from the equator towards either pole it

[1] A few pages in this chapter are extracted from a paper by the author in the *Proceedings of the Birmingham Philosophical Society*, vol. IX. part I. 1893.

gradually gets heavier, and the pull is about 1 in 200 greater near the poles than at the equator. An ordinary balance, used in the ordinary way, will not show this change, for it equally affects the contents of either pan. It would undoubtedly be shown by a spring balance if we could only get a spring at the same time sensitive and constant in its action. But springs are in general by no means constant or consistent. They have, as it were, memories. They remember any change in stretch and any change in temperature to which they have been subjected, so that after a change and a return to the original external conditions their action is not quite what it was before. It is true that their memory fades, but not sufficiently to let us make quite consistent weighings. There is only one kind of solid spring known which has no appreciable memory, one made of quartz fibre. With such a spring Mr Threlfall has succeeded in showing change of weight with change of place.

Since, then, an ordinary balance fails, and a spring balance is too inconsistent, to show the change in earth-pull, how do we know that the change exists and even what it amounts to? Fortunately we have an excellent detector of weight change in the pendulum. If a pendulum like that of a clock is supported so that it can swing quite freely to and fro, the time that it takes to make one swing depends on its shape and size and on the pull of the Earth downwards on its bob. If the pull on the same pendulum increases, the time of swing decreases at half the rate. Now the time of a swing can be measured with very great accuracy, for we can watch the pendulum for hours and count the number of swings in the total time of watching. Dividing the total time by the number of swings we get the time of one swing.

Pendulums have been carried about the world and the times of swing of the same pendulum have been exactly measured in widely different latitudes. The results of these measurements show quite conclusively that the weight of the bob of a given pendulum increases as we travel polewards from the equator, and we may thus describe the change. If we had a perfect pendulum clock compensated for temperature change and barometer change (for if the density of the air changes, so does the effect on the buoyancy of the pendulum change), then on removal from the equator to this country it would gain about 130 seconds a day, and on removal from the equator to the pole it would gain about 216 seconds a day.

There is a change in the weight of a body not only if we remove it north or south on the level but also if we change its level by raising it in a vertical line. Assuming that the pull on a body above the Earth's surface is inversely as the square of its distance from the Earth's centre, the weight of a body should decrease about 1 in 2000 for a rise of 1 mile or by about 1 in 10 millions for a rise of 1 foot. If, then, we have a balance as in fig. 22, p. 74, with two sets of pans, P and Q at one level, and P' and Q' at another lower level, and if the weights A and B exactly balance against each other at the PQ level, they will still exactly balance against each other if they are both removed to the lower $P'Q'$ level, for each gains in weight in the same proportion. But if while B is left in the upper pan at Q, A is taken out of P and put into P', it alone gains in weight and the balance will tilt down a little on the P side.

Several experiments were made in the 17th and 18th centuries to look for this change of weight and to show it by the balance, but it was first detected and measured by von Jolly at Munich about 1878. He set up a balance with the pans PQ at the top of a tower and with the lower pans $P'Q'$ 21 metres—say 23 yards—below. He balanced two 5 kilogramme weights against each other at the top. Then the weight in P was removed to P' and the gain in weight was 32 milligrammes or about 64 in 10 millions, rather less than 69 in 10 millions given by the inverse square law for the 69 feet change in level. We shall describe later the device by which the difference in air buoyancy at the two levels was eliminated.

Some years later Richarz and Krigar-Menzel succeeded in measuring the change in weight with a change in level of only 2·3 metres—say 2½ yards. They used a kilogramme weight in each pan, and on moving a kilogramme on one side from the upper to the lower level it gained about 0·65 milligrammes or 6·5 in 10 millions, whereas the gain according to the inverse square law should have been about 7·5 in 10 millions. In each case the increase was less than according to the law, probably through the attraction of the surrounding building or neighbouring elevated ground. The law, indeed, could only be expected to hold over the surface of the ocean.

These experiments with the pendulum and with the balance show us conclusively that the weight of a given piece of matter—the earth-pull on it—varies with its situation. But there is a property or quality which remains the same for the same matter everywhere and always. This quality is its inertia or its mass. And the idea underlying inertia is the effort required to get up a certain speed in the body. If a greater effort is needed to get up the speed in one body than in another, the first body has the greater inertia. We give quantitative expression to the idea by saying that an equal force is required everywhere and always to give the same rate of gain of speed in the same piece of matter, and we say that it always has the same mass. If there are two bodies and we have to put double the force on to one that we have to put on to the other for the same rate of gain, the first has double the mass of the second, or generally the mass of a body is proportional to the force needed to produce a given rate of gain of speed[1].

It is not easy to make exact and direct experiments to test the constancy of a given piece of matter, and the difficulty lies in applying equal forces at different places. But the experiment is being made for us continually, in a rather complicated form, by ships' chronometers. The rate of a chronometer is decided by the vibration of the balance wheel against the coiling and uncoiling of the hair spring and the weight of the wheel does not come into account. We must suppose that, at the same temperature, the spring offers the same resistance to the same coiling wherever it may be, and as the chronometer keeps the same time at the same temperature in all latitudes, the rate of change in speed of the balance wheel must be the same in a given part of its vibration wherever it may be. In other words the rate of change of speed under a given force is everywhere the same, or the mass of the wheel is constant. It may be noted that

---

[1] It is not necessary to take into account here the interpretation of certain recent experiments as implying a change of mass when a body is made to change its speed. Such change could not be appreciable unless the speeds were enormously greater than any that we are considering.

we do not here suppose that we use the chronometer at different temperatures. Our argument would fail if we did so. For the resistance to coiling of the spring changes with change of temperature, and the chief aim of compensation is to correct for this change.

A similar experiment might be made with a tuning fork. Here again the time of vibration depends on the resistance of the prongs to bending in or out and only in quite negligible degree on their weight if they are always used in the same position. Let us suppose that a fork is tested at the same temperature in two different latitudes under conditions which give us reason to suppose that it is the same fork and not altered by rust or wear, and at such small interval of time that we are entitled to take its resistance to bending as unaltered. Then if we find the number of vibrations per second we have the same rate of change of motion under the same force, or the mass is constant. Though this experiment has never been made deliberately for this purpose, it has, no doubt, often been made in the verification of the vibration-frequency marked on forks used in laboratories, and the constancy of mass is verified to the same order of accuracy.

Our general conclusion from observation is that the rate of change of motion of a body when controlled by its weight changes with the place of observation, while the rate of change when controlled by a spring, which we may fairly consider to have constant properties, is the same everywhere. Hence it is the weight of a body that varies and not its mass, a conclusion which has been taken as true for nearly 250 years and has never led to the least inconsistency.

Newton was the first to make the idea of mass definite, and he showed that *so long as we are at the same place* the weight or the earth-pull on bodies is exactly proportioned to their masses. If we have one body twice the weight of another, then whatever the bodies are, say one gold, the other wood, the first has exactly twice the mass of the other. This could be roughly verified by repeating Galileo's famous experiment, in which he dropped at the same instant various bodies over the edge of the Leaning Tower at Pisa and showed that they reached the ground at the same time. This meant that every pound of weight had the same amount of mass to pull on, whatever body the mass belonged to. In Galileo's experiment the resistance of the air came in to interfere with exactness. Newton used a much more accurate test. He made two pendulums, each consisting of a hollow round box, and these were hung by strings 11 feet long so that they might vibrate side by side. Into the boxes he put equal weights of different substances, such as gold, silver, lead, glass, sand, salt, wood, wheat, and he found that the two pendulums, if started together, continued to swing together for a long time. The air resistance was the same and the wood boxes were the same for both. The only difference was the kind of matter inside the boxes, and as the equal pulls produced equal changes of speed for quite a long time, the masses of the different equal weights must have been equal. If there had been a difference, if, for example, the gold had more mass than wood of the same weight, the gold would have taken a longer time for each vibration than the wood, and the two would have got more and more out of step. The effect being thus cumulative, would in the long run have shown even a very small difference in mass.

Since weight is thus shown to be exactly proportional to mass, when the weighing is carried out at the same place, we may use the balance to weigh out different masses, and, indeed, this is precisely what the balance does for us. The qualities of bodies for which we purchase them are in proportion to their mass and not to their weight. A lump of sugar has the same sweetness here and at the equator, though it is heavier here. A ton of coal has the same heating power in either region, though its weight is greater here by an amount equal to the earth-pull on six pounds. Our ordinary description of the pieces of iron or brass we use on a balance as being 'weights' does not tend to clear thought on this point. What we term a 'pound-weight' is really a 'pound-mass,' and is the same wherever it may be carried about the world. The weight of that pound varies from place to place, and we have to remember that the weight of a pound and a pound-weight involve different ideas.

We can now imagine an experiment which would give us the mass of the Earth by direct weighing—if only it could be carried out. Let us suppose that we could divide the whole Earth into blocks, each, say, a cubic foot in size. Let one of the blocks be brought up to a certain place, weighed there, and then put back. Then let another of the blocks be brought to the same place, weighed, and put back, and so on until every block has been weighed. The sum of all the 'weights' is really the sum of the masses or is the mass of the Earth.

The experiments which we shall describe later show that the result of such weighings would be about 13·2 million million million million pounds or $13·2 \times 10^{24}$ lbs., a number so vast that we attach no idea to it beyond its vastness. But the mass of the Earth is expressed in a more thinkable way in terms of the mass of an equal volume of water. At the rate of 62·4 lbs. per cubic foot this would be about $2·4 \times 10^{24}$ lbs. Thus the average density of the Earth is about $13\cdot2 \times 10^{24} \div 2\cdot4 \times 10^{24} = 5\frac{1}{2}$ times that of water. Taking the density of water as 1 the result is that the Mean Density of the Earth is about $5\frac{1}{2}$, and this is the way in which its mass is always expressed.

Though the imagined experiment would be exactly and truly an Earth-weighing experiment, it can only be imagined. We can make no approach to carrying it out in practice. Our deepest mines reach down hardly a mile, so that we make only slight scratches on the surface, and know nothing directly of the deeper layers.

We require, then, to measure the Earth's mass, some other property of matter than mere earth-pull on it and such a property was discovered by Newton when he showed that a piece of matter is pulled not only by the Earth but by every other piece of matter in proportion to the mass of either piece and inversely as the square of their distance apart. Or, the pull of a mass A on a mass B distant d from it, is proportional to

$$\frac{\text{Mass of } A \times \text{Mass of } B}{d^2},$$

which is Newton's Law of Gravitation.

Newton showed that a sphere such as the Earth, with density the same all round at the same distance from the centre, will pull on any outside body just as if all the mass of the sphere were collected into one single point at the centre.

Now consider a body supported just above the surface of the Earth 4000 miles from the centre. We know that the pull on it will make it fall 16 feet in the first second if it is allowed to drop. If we could take it up 4000 miles, or twice as far from the centre, and then let it drop, the law says it would fall $\frac{1}{2^2}$ or $\frac{1}{4}$ of 16 feet in the first second. If we could take it up 12,000 miles from the centre it would fall $\frac{1}{3^2}$ or $\frac{1}{9}$ of 16 feet, and so on. So that if we could take it to 60 times the distance of the surface from the centre or 240,000 miles it would fall $\frac{1}{60^2}$ or $\frac{1}{3600}$ of 16 feet or just about $\frac{1}{13}$ inch in the first second.

It is just at this distance that we have a body by which we can test the law. The moon is moving nearly in a circle round the Earth's centre, and with a velocity about 3400 feet per second. Let A, fig. 13, be the position of the Moon's centre at the beginning and B its position 3400 feet further along the curve at the end of a particular second. Were it moving at A free from the pull of the Earth it would move to T, along the tangent at A, where AT is 3400 feet.

TB is the distance it drops in the second and it is easy to show that *TB* is very nearly $\frac{1}{19}$ inch or is what we expected from the law. Hence every pound of the Moon's mass is pulled by a force $\frac{1}{3600}$ of the pull on an equal mass at the Earth's surface.

Comparing the motions of the different planets under the pull of the Sun, it can be shown that with them also the pull in each case is proportioned to the mass of the planet and to the inverse square of its distance from the Sun. In fact a pound of mass has

Fig. 13.

on it a pull by the Sun inversely as the square of its distance from the Sun's centre whatever the planet of which it forms a part. So the law is amply verified as regards the mass pulled and the distance.

To show that the pull is also proportional to the mass of the pulling body, we assume the law, which holds good in all cases which we investigate, that if two bodies *A* and *B* act on each other, the force which *A* exerts on *B* is equal though opposite to the

P.   4 force which B exerts on A. We shall, for simplicity, neglect the difference of distance of different parts of the Earth and Moon from each other. As each body pulls the other as if it were concentrated at its centre, this simplification is justified. The Earth pulls the Moon with a force proportional to the mass of the Moon, so that each pound in the Moon is pulled with an equal force. In turn each pound pulls the Earth with an equal force and the total is proportional to the number of pounds of mass pulling. Thus we may conclude that Newton's statement holds good that the gravitation pull is proportional to the product of the masses of the two pulling bodies.

Now we can see how the gravitative pulls of two bodies on a third body enable us to compare their masses. Let A and B, fig. 14, be two bodies and let us suppose that we wish to determine the mass of A in terms of the mass of B. Let a third body m be distant a from A and b from B and let A pull it with force α and B pull it with force β.

The law of gravitation gives us

$$\frac{\alpha}{\beta} = \frac{\dfrac{\text{Mass of } A \times \text{Mass of } m}{a^2}}{\dfrac{\text{Mass of } B \times \text{Mass of } m}{b^2}} = \frac{b^2}{a^2} \frac{\text{Mass of } A}{\text{Mass of } B},$$

whence $\quad$ Mass of $A = \dfrac{a^2}{b^2} \cdot \dfrac{\alpha}{\beta}$. Mass of B.

We can for example find the mass of the Sun in terms of the mass of the Earth. If we suppose A, B, and m to be, respectively, the Sun, the Earth, and the Moon, then a is 93 million miles and b is 240,000 miles. The ratio of the pulls α/β in the two circles, one described in a year and the other in $\frac{1}{13}$ of a year, can be calculated and it works out to be about $\frac{120}{64}$. Approximately then

Mass of Sun = 300,000 Mass of Earth.

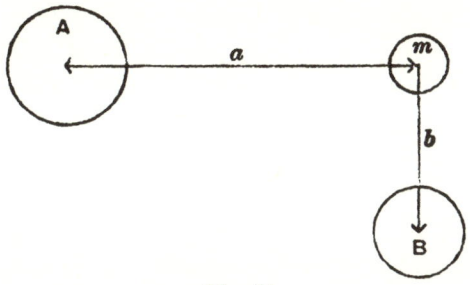

Fig. 14.

But for the purpose of Earth weighing, A must be the Earth, while B must be some body of which we know the mass in pounds or kilogrammes, and we must be able to find what is the ratio of the pulls of the two on a third body m.

Newton discussed the possibility of comparing two such pulls, and in two ways. In one of these he thought of comparing the attraction of a mountain

with that of the Earth. If a plumb·bob were hung
at the side of the mountain, the mountain would
draw the bob towards it. If AC (fig. 15) is the
direction in which it would hang if the mountain
were removed and if AB is its actual direction it is
easily seen that, if we consider only the horizontal
part of the mountain-pull,

$$\frac{\text{Mountain-pull}}{\text{Earth-pull}} = \frac{CB}{AC}.$$

If then we can measure the angle of deflection of the

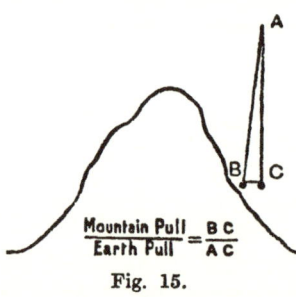

Fig. 15.

plumb line BAC we can
determine CB/AC and
therefore the ratio of the
pulls. Newton calculated
that if the mountain were
hemispherical, 3 miles
high, and of the same
density as the Earth, a
plumb bob at its base
would not be deflected
so much as 2 minutes. In fact, as the two attractions would be in the ratio of radius of the mountain
to the diameter of the Earth, a bob with a string a
yard long would be drawn aside $\frac{3}{8000} \times 36$ inch or $\frac{1}{74}$
inch. If the actual drawing aside were $\frac{1}{2}$ or $\frac{1}{4}$ of this
we should then know that the density of the moun-
tain was $\frac{1}{2}$ or $\frac{1}{4}$ that of the Earth.

In the other way of comparing pulls Newton considered the possibility of using a sphere for B, fig. 14, and another sphere for the mass m and he calculated that two spheres of the density of the Earth and each a foot in diameter, if to begin with they were $\frac{1}{4}$ inch apart, would take not less than one month to draw together into contact. There was a mistake in arithmetic here, for the time would really only be about 320 seconds.

Newton dismissed the subject with the remark that in neither case would there be an effect great enough to be perceived—a statement no doubt true for the methods of measurement then available. But the enormous extension of scientific theory, so largely due to Newton, was accompanied by a great improvement in the methods of measurement, and what to him seemed impossible, was actually tried about ten years after his death by Bouguer.

Bouguer was a member of the expedition sent out by the French Academy to measure, as described in the last chapter, the length of a degree of latitude at the Equator, and, impressed by the vastness of the Andes, he determined to try to measure the ratio of the pull of a mountain on a plumb bob to the pull of the Earth. For his purpose he fixed upon Chimborazo, a mountain some 20,000 feet high, as most suitable, and he selected a station on the south slope just above the snow-line and about 5000 feet below the summit. Here he and his colleague, de la Condamine, fixed their tent after a most toilsome journey of ten hours over rocks and snow, and in face of great difficulties due to frost and snow, they took the zenith distances of several stars as these crossed the meridian. Then a few days later they moved to a second station very nearly four miles west of the first, where the attraction of the mountain had only a small component towards the north, not more than $\frac{1}{12}$ the value it had at the first station. Here their difficulties were even greater than before. They were exposed to the full force of the wind which filled their eyes with sand and was continually on the point of carrying away their tent. The cold was intense, and so hindered the working of their instruments that they had to apply fire to the levelling screws before they could turn them. Still, they made their observations, measuring the distances from the zenith of the same stars as they crossed the meridian. The principle of the method may be seen from fig. 16, where we suppose that the stars are looked at through a telescope provided with a plumb line hanging from its upper end. Imagine that we begin at the second station represented in the lower figure, and watch the passage of a star which for simplicity we will suppose to cross the meridian exactly at the zenith. Let us suppose that at the first station the vertical

is deflected by the mountain. Then the same star
will appear at that station to be displaced from the
zenith towards the north. The average for different
stars was found to be about $7\frac{1}{2}$ seconds. Making
corrections for the small deflection towards the north
at the second station Bouguer estimated the deflection of the plumb
line at the first station
to be about 8 seconds.
Had Chimborazo been
of the density of Earth,
Bouguer calculated that
it would have drawn the
vertical aside about 12
times as much, or the
Earth appeared to have
a density 12 times that
of the mountain, a re-
sult undoubtedly far too
large. But it is little
wonder that under such
adverse circumstances

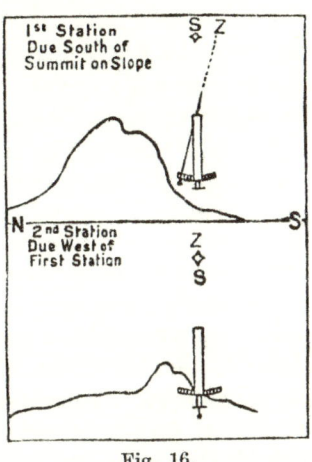

Fig. 16.

the experiment failed to give a good result. Not-
withstanding the failure, both in this experiment
and in another which we shall not describe, great
honour is due to Bouguer in that he showed that
Earth-weighing is possible. He showed that mountains do really attract, and that the Earth, as a

whole, is denser than the surface strata. As he remarked, his experiments at any rate proved that the Earth was not merely a hollow shell, as some had till then held; nor was it a globe full of water, as others had maintained. He fully recognised that his experiments were mere trials, and hoped that they would be repeated in Europe.

Thirty years later his hope was fulfilled. Maskelyne, then the English Astronomer Royal, brought the subject before the Royal Society in 1772, and obtained the appointment of a committee 'to consider of a proper hill whereon to try the experiment, and to prepare everything necessary for carrying the design into execution.' Cavendish, who was himself to carry out an Earth-weighing experiment some twenty-five years later, was probably a member of the committee, and was certainly deeply interested in the subject, for among his papers have been found calculations with regard to Skiddaw, one of several English hills at first considered. Ultimately, however, the committee decided in favour of Schiehallion, a mountain near Loch Rannoch, in Perthshire, 3,547 feet high. Here the astronomical part of the experiment was carried out in 1774, and the survey of the district in that and the two following years. The mountain has a short east and west ridge, and slopes down steeply on the north and south, a shape very suitable for the purpose.

Maskelyne, who himself undertook the astronomical work, decided to work in a way very like that followed by Bouguer on Chimborazo, but modified in a manner which Bouguer had suggested. Two stations were selected, one on the south, and the other on the north slope. A small observatory was erected, first at the south station, and the angular distance of some stars from the zenith, when they were due south, was most carefully measured. The stars selected all passed nearly overhead, so that the angles measured were very small. The instrument used was the zenith sector, a telescope rotating about a horizontal east and west axis at the object glass end, and provided with a plumb line hanging from the axis over a graduated scale at the eye-piece end. This showed how far the telescope was from the vertical when it was directed to a star not overhead.

After about a month's work at this station, the observatory was moved to the north station and again the same stars were observed with the zenith sector. Another month's work completed this part of the experiment. Fig. 17 will show how the observations gave the attraction due to the hill. Let us for the moment leave out of account the curvature of the Earth, and suppose it flat. Further, let us suppose that a star is being observed which would be directly overhead if no mountain existed. Then evidently at S. the plumb line is pulled to the north, and the

zenith is shifted to the south. The star therefore
appears slightly to the north. At N. there is an
opposite effect, for the mountain pulls the plumb
line southwards, and shifts the zenith to the north;
and now the same star appears slightly to the south.
The total shifting of the star is double the deflection
of the plumb line at either station due to the pull of
the mountain.

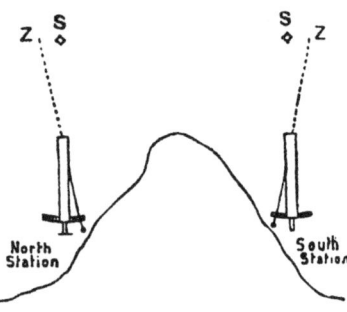

Fig. 17.

But the curvature of the Earth also deflects the
verticals at N. and S., and in the same way, so that
the observed shift of the star is partly due to the
mountain, and partly due to the curvature of the
Earth. A careful measure was made of the distance
between the two stations, and this gave the curvature
deflection as about 43″. The observed deflection was

about 55″, so that the effect of the mountain, the difference between these, was about 12″.

The next thing was to find the form of the mountain. This was before the days of the Ordnance Survey, so that a careful survey of the district was needed. When this was complete, contour maps were made, and these gave the volume and distance of every part of the mountain from each station. Hutton was associated with Maskelyne in this part of the work, and he carried out all the calculations based upon it, being much assisted by valuable suggestions from Cavendish.

Now had the mountain had the same density as the Earth, it was calculated from its shape and distance that it should have deflected the plumb lines towards each other through a total angle of 20·9″, or $1\frac{4}{5}$ times the observed amount. The Earth, then, is $1\frac{4}{5}$ times as dense as the mountain. From pieces of the rock of which the mountain is composed, its density was estimated as $2\frac{1}{2}$ times that of water. The Earth should have, therefore, density $1\frac{4}{5} \times 2\frac{1}{2}$ or $4\frac{1}{2}$. An estimate of the density of the mountain, based on a survey made thirty years later, brought the result up to 5. All subsequent work has shown that this number is not very far from the truth.

An exactly similar experiment was made eighty years later, on the completion of the Ordnance Survey of the kingdom. Certain anomalies in the direction

of the vertical at Edinburgh led Colonel James, the
director, to repeat the Schiehallion experiment,
using Arthur's Seat as the deflecting mountain. The
value obtained for the mean density of the Earth was
about 5⅓.

Experiments have also been made in which the
attraction of a part of the Earth's crust such as a
mountain, or the layers above the bottom of a mine,
has been compared with that of the whole Earth
by its effect in altering the time of vibration of a
pendulum. This method was employed in Bouguer's
second experiment mentioned above. But it has
never yielded satisfactory results. Indeed it is now
recognised that, in common with the method of the
deflection of the vertical by a mountain, it is not very
trustworthy. For in the first place there is inevitable uncertainty in the density of the part of the
crust used. Even if we knew the density of Schiehallion exactly, there is ignorance of the density of the
strata underneath. Often there appears to be a
defect in the attraction which might be expected to
arise from tablelands and mountain ranges, and Airy
made a suggestion that these raised masses may be
buoyed up, like the peaks of icebergs, by lighter
matter below. In cases of very ancient and sinking
rocks there may be heavier matter below. In the
second place, in calculating the effect of a mountain
we must take into account the attraction of other

raised matter in the neighbourhood and it is a question how far we are to go. Hutton in the Schiehallion experiment stopped at 3 miles. But a mass eight times as great at 6 miles would have an equal disturbing effect and any large raised mass at the greater distance should be taken into account. For these reasons probably, the results of the various experiments in which a 'Natural Mass' has been used, such as a mountain or the Earth's upper strata, have varied over a considerable range.

We turn now to the second method of experiment considered by Newton, in which is measured the attraction between two spheres, each of known size and at a known distance from centre to centre. This we may call the 'Prepared Mass' method. Let us suppose that we find that a sphere of mass M attracts another sphere of mass m with a force P when their centres are d apart and that the Earth of mass E and radius R attracts m with force W, its weight. Assuming that the Earth attracts as if it were all collected at its centre we have

$$\frac{E}{R^2} : \frac{M}{d^2} = W : P.$$

Then
$$E = M\,\frac{WR^2}{Pd^2}.$$

We have then to find the pull P at distance d due to the mass M on a sphere of weight W.

The idea of making such an experiment occurred towards the end of the 18th century to the Rev. John Michell, the discoverer of the inverse square law of magnetic action. The 'torsion balance' for the measurement of forces such as those between magnetic poles was invented independently by Michell and by Coulomb. It consists of a horizontal rod suspended from its centre by a thin wire or fibre which resists a twist. The force to be measured is then applied at one end of the rod in a horizontal direction and at right angles to the rod, and the rod is pulled or pushed round by the force. If the force is very small, the angle of twist is usually small, and its measurement enables us to find the force. Michell saw the possibility of measuring the gravitative pull between masses not too large to handle or move, and constructed some apparatus for the purpose. He died in 1793 without making any experiments with it and after his death the apparatus came into the hands of Cavendish, the great chemist and physicist, one of whose achievements was the discovery of the constitution of water.

Cavendish reconstructed most of the apparatus, and in the years 1797–8 he carried out the great Earth-weighing experiment known as the Cavendish experiment. Though the idea was due to Michell it is right that Cavendish's name should be attached to the work, for the details both of the apparatus and of the mode of using it are due to him, and he made the experiment in a manner so admirable that it marks the beginning of a new era in the measurement of small forces.

Cavendish sought to measure the pull between a lead sphere 12 inches in diameter weighing about

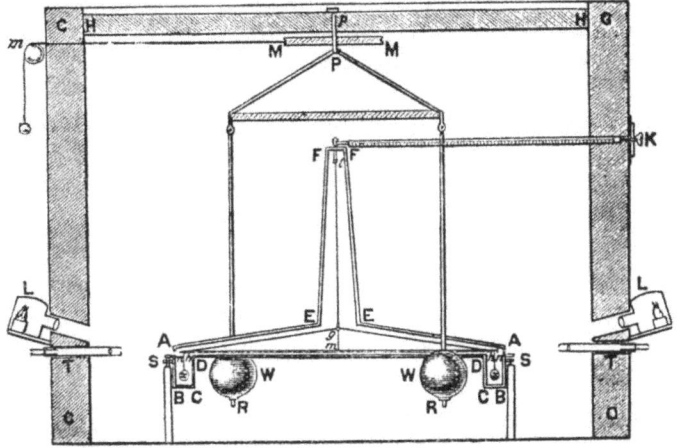

Fig. 18. Cavendish's Apparatus. Elevation.

hh, torsion rod. *xx*, balls hung from its ends. *WW*, attracting masses movable round axis *P*. *TT*, telescopes to view position of torsion rod.

350 lbs., on a lead sphere 2 inches in diameter and weighing about 1 lb. 10 ozs. when the distance between their centres was about 9 inches. The apparatus is represented in elevation in fig. 18 and in plan in fig. 19. It was enclosed in a chamber *GGGG* built within

another to ward off changes of temperature, and the
air currents thereby produced. The torsion rod *hh*
was of deal 6 feet long tied by wires *hg* to an upright
mg to give strength and rigidity. In order to double
the effect there were two attracted 2 inch spheres *xx*
hung by short wires from the ends of the rod and the
rod itself was hung by a wire *lg* of silvered copper
about 40 inches long from the top of the protecting
case at *F*. There were also two attracting spheres
WW, each 12 inches in diameter, hung from a cross
piece as shown in fig. 18 and these could be moved

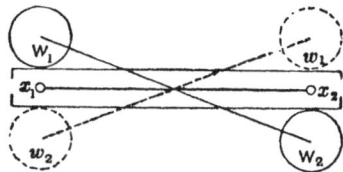

Fig. 19. Cavendish's Apparatus. Plan.
Attracted balls x_1x_2. Attracting masses W_1W_2.

from the positions W_1W_2 in fig. 19 to the positions
w_1w_2 round an axis coinciding with the axis of the
wire *lg*, by a cord passing outside the enclosing
chamber at *m*.

The position of the torsion rod was determined by
a mark (really a vernier) on the end of the rod which
moved over a divided scale fixed near the end. The
scale was lighted by a lamp *L* and viewed by a
telescope *T*.

In fig. 18 the two attracting spheres WW are not in position for exercising the maximum pulls on xx. They would have to be moved round a little further to give the positions W_1W_2 of fig. 19. They were stopped in the latter position by pieces of wood when $\frac{1}{8}$ inch from the case and with just under 9 inches from centre of W to centre of x.

The apparatus was to give P, the pull of W on x at 9 inches. Imagine that we begin with the spheres WW far away (or, what is equivalent, in the line at right angles to the torsion rod through its centre) and that we read the position of the end of the rod on the scale. Now bring the masses into the positions W_1W_2, fig. 19, when there is a pull P at each end of the rod, turning the rod round, and we observe that it turns through n divisions of the scale. Then move the masses round to the positions w_1w_2 when the two pulls P are reversed and the rod moves round through n divisions from its first position in the opposite direction. The total change of reading of the rod on the scale between the W_1W_2 and the w_1w_2 positions of the masses will be $2n$ and it will be unnecessary to observe the reading when the masses are half-way between, the equivalent to being far away. The deflection of $2n$ divisions is equal to the deflection which would be produced by $4P$ applied at one end only.

The next thing is to find the actual force

corresponding to the observed number of divisions on the scale. When a system of this kind is suspended by a wire so that it can vibrate in a horizontal plane, twisting and untwisting the wire, the time of one vibration to and fro is the same whatever the extent of the excursion and depends on the arrangement of the mass of the system about the axis of vibration and on the force, applied at the end of the arm, needed to twist the wire through unit angle. If then we observe the time of vibration and know how the mass is disposed, we can find the force which will twist the system through the unit angle. But the force for any other angle is in proportion to the angle, so that we can calculate the force $4P$ needed to twist through $2n$ divisions, or the force P twisting through $\frac{1}{2}n$ divisions.

We may put this into simple mathematical form if we neglect all corrections. Suppose that the $2n$ divisions correspond to an angle of deflection θ, and that the torque per radian twist of the end of the wire is μ. Then if a is the arm at which P acts,

$$4Pa = \mu\theta \ \dots\dots\dots\dots\dots\dots(1).$$

If I is the moment of inertia of the vibrating system round the axis of the wire and if T is the time of one vibration,

$$\mu = \frac{4\pi^2 I}{T^2} \ \dots\dots\dots\dots\dots(2),$$

whence $$P = \frac{\pi^2 I \theta}{T^2 a},$$

and going back to the formula on p. 61, in which M now represents the mass of one of the attracting spheres WW (fig. 18), W is the weight of one of the attracted spheres xx, and d is the distance of the centres apart,

$$E = \frac{MWR^2}{Pd^2} = \frac{MWR^2 T^2 a}{\pi^2 I \theta d^2} \quad \ldots\ldots\ldots\ldots (3).$$

As we now know the quantities on the right hand of (3), we have determined E the mass of the Earth.

We have supposed in this account that only the masses W acted on the masses x, but in reality the rods suspending the masses exercised some attraction, and both masses and rods exercised some attraction on the torsion rod hh. Further, each mass was attracting not only the ball nearest to it but also to a small extent the further ball, and all these attractions had to be taken into account and allowed for, the observed value of P being the value used with formula (1) multiplied by a certain factor which could be determined from the arrangement and dimensions of the apparatus.

Cavendish made 29 separate determinations, and the value for the mean density of the Earth resulting from these determinations is 5·448. This is corrected for a mistake which was detected in the original paper many years after its publication.

The experiment has been repeated several times since by other workers and the most notable repetition is that by Professor C. V. Boys, who published an account of his experiment in 1895.

Boys had a few years before invented a method of drawing out fibres of quartz of great fineness—a diameter of $\frac{1}{10000}$ inch being quite easily obtained. He found that these fibres are extraordinarily strong for their diameter and extraordinarily true in their elastic properties. A quartz fibre may be twisted round very many turns and, on being released, it will untwist the same number of turns and come back, as nearly as can be determined, to its original position ; whereas a metal wire thus twisted acquires 'permanent set,' and on release does not untwist the whole way back to the original position. By this great invention Boys put into the hands of physicists a means of making torsion balances for the measurement of small forces far exceeding in delicacy and accuracy anything hitherto used. He determined to repeat the Cavendish experiment, using a quartz fibre instead of a metal wire to suspend the torsion rod. It was necessary to reduce the size of the vibrating system to be small enough to be carried by a fine fibre. This reduced the sizes and distances to be measured and it was perhaps more difficult to measure these sizes and distances with proportionate accuracy ; but, on the other hand, the twist for a small force increased with the finer fibre and the apparatus

became so small that it could be kept at a much more
uniform temperature, and air currents, which are in
a closed case entirely due to uneven temperature at
different parts of the case, were thereby very greatly
reduced.   These air currents are the chief disturbers
in such an experiment and were found to be very
troublesome in the larger apparatus used by
Cavendish.

The suspending fibre which Boys used was about
17 inches long, and probably about $\frac{1}{5000}$ inch in diameter.   The torsion rod was only $\frac{9}{10}$ inch long in place
of Cavendish's 6 foot rod.   The attracted spheres at
its ends were in one set of experiments gold balls
$\frac{1}{4}$ inch in diameter, and the attracting spheres were
lead $4\frac{1}{2}$ inches in diameter.   The torsion rod was
itself a mirror and the image of a divided scale
22 feet away was viewed in the mirror by a telescope.

If the attracting and attracted masses had all
been on one level as in Cavendish's experiment, it
will be seen from the plan in fig. 21 that with a
distance of less than an inch between the attracted
masses a $4\frac{1}{2}$ inch sphere in front of one mass would
have been almost equally in front of the other and
with nearly the same distance between centres, and
so, pulling them almost equally in the same direction,
would not have tended to turn the rod round much.
Boys therefore adopted the plan represented in elevation in fig. 20.   The attracted masses were suspended

by quartz fibres from the ends of the mirror torsion rod at different levels, one 6 inches below the other, and one of the attracting balls was on each of these levels.

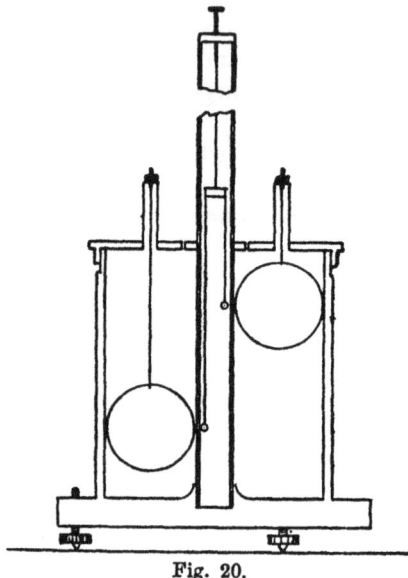

Fig. 20.

The attracted balls hung in an inner protecting tube and the attracting balls hung in an outer case which surrounded the inner tube and could be revolved round it. Fig. 21 represents a plan on which the centres m_2 and M_2 must be supposed to be 6 inches

below the plane containing the centres of m_1 and
M_1. As the case containing the attracting balls was
revolved there was a position M_1M_2 in which the
moment of the pulls on the attracted balls m_1m_2 was
a maximum in one direction and a position $M_1'M_2'$
in which it was a maximum in the other direction.
Thus each attracting mass acted in both its positions
on the same attracted mass. The general theory of

the experiment is like
that of the Cavendish
experiment and we need
not repeat it. The final
result of Professor
Boys's work gave the
mean density of the
Earth as 5·527, and for
the present this may be
taken as the most trust-
worthy result.

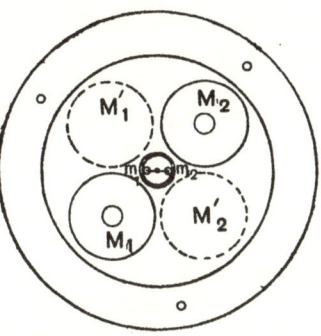

Fig. 21.

We have now to de-
scribe another mode of
experiment, in which the pull between two masses is
measured by the common balance instead of by the
torsion balance. Though the common balance is in
some ways less satisfactory for the purpose, it is
well in work of this kind, where the quantity to be
measured is small, to have different modes of attack.
For there might possibly be some undetected error,

characteristic of one method, which a divergence of result by another method would reveal. An agreement by the two methods gives us confidence in both.

The first account of a common balance experiment was published by the late Professor von Jolly of Munich in 1878. In his final work, a little later, a balance was mounted on a support at the top of a tower, with scale pans under the two ends of the beam in the usual position. Another pair of scale pans was suspended by wires from these, 21 metres, say 23 yards, below, nearly at the bottom of the tower as represented in fig. 22. Four glass globes *ABCD* of equal weights and volumes were prepared and two of them, *A* and *B*, were filled each with 5 kgm. of mercury. Then all four were sealed. First *A* and *B* were put in the upper pair of pans, and *C* and *D* in the lower pair, and a balance was made. Then *A* and *C* were interchanged. The equality of volume of the two globes eliminated any effect due to the greater buoyancy of the air below and there was a gain in weight rather more than 31 mgm., due to the approach of the 5 kgm. of mercury to the Earth. This is the first experiment in which a change in the weight of a body in so small a change in height as 21 metres was demonstrated.

A lead sphere about 1 metre in diameter was now built up out of separate blocks, immediately under

one of the lower pans. On again effecting the interchange between *A* and *C*, *A* when brought below
weighed 0·59 mgm. more than it did before, and this
was the pull on it by the lead sphere. The distance

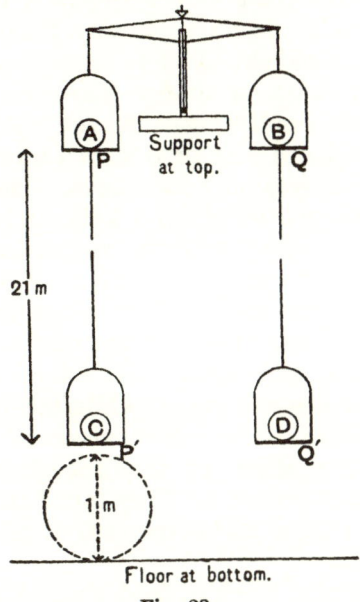

Fig. 22.

from centre of lead to centre of mercury was about
57 cm. If, then, the lead sphere at an effective
distance of 57 cm. exercised a pull of 0·59 mgm. on

the mercury, and the Earth at an effective distance equal to its radius, about 690 million centimetres, exercised a pull of 5 kgm. or 5 million milligrammes, the mass of the Earth could at once be calculated in terms of the mass of the sphere of lead. When the result was put in the usual way, the mean density of the Earth came out as 5·69.

An experiment on similar lines was carried out later by Richarz and Krigar-Menzel. Like von Jolly, they had a balance with pans at two levels, but their change in level was only 2·3 metres. They used two solid spheres each weighing 1 kgm. and two hollow spheres of the same external volume as these and weighing 53 gm. each. Virtually they began with a solid sphere above and a hollow sphere below, on the left say, and the reverse arrangement on the right. Then on each side solid and hollow were interchanged, and the left gained while the right lost by the interchange. The effect observed was therefore twice the effect of the change in level of $1000 - 53 = 947$ grammes. They found that the effect of lowering 1 kgm. 2·3 metres was a gain in weight of 0·65 mgm.

A rectangular block of lead about 2 metres high and nearly cubical was then built up of separate pieces under the balance and between the two levels. There were narrow vertical tunnels through the block for the passage of the wires to which the lower pans were attached. When, starting with solid above and hollow below on the left and with the reverse on the right, an interchange was made of solid and hollow, the left-hand solid had the attraction of the lead changed from a pull down to a pull up, while the right hand had the reverse change. The effect of the interchange was therefore that of change in height minus four times the pull of the lead block on one sphere. The experiment gave the attraction of the lead block on one sphere as 0·36 mgm., whence the mass of the Earth could be found in terms of the mass of the lead. The mean density of the Earth deduced was 5·505.

About the same time that von Jolly began his experiment the author also saw the possibility of using the common balance to measure the attraction between two masses and made some preliminary trials which ultimately led to an experiment, which was carried out at Birmingham. As the author knows more about this experiment than about the other experiments by the common balance, it is selected for more detailed description.

The balance (fig. 23) was of the type used at mints to weigh out bullion. It had a specially strong beam 4 feet long. It was supported on two iron girders, seen in section in *gg*, and these were supported on two brick pillars, of which the one at the back only is shown. In order to prevent the vibrations due

to street traffic and to the shutting of doors in the
building one course of brickwork in each pillar was
replaced by a number of indiarubber blocks. The

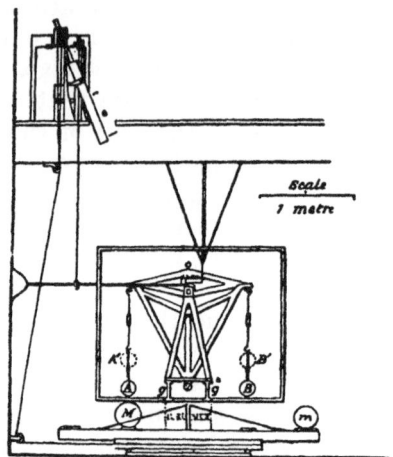

Fig. 23.

AA, weights, each about 50 lbs., hanging from the two arms of
balance. *M*, attracting mass on turn-table, movable so as to
come under either *A* or *B*. *m*, balancing mass. *A'B'*, second
positions for *A* and *B*. In these positions the attraction of *M* on
the beam and suspending wires is the same as before, so that
the difference of attraction on *A* and *B* in the two positions is
due to the difference in distance of *A* and *B* only, and thus the
attraction on the beam, &c., is eliminated.

balance was enclosed in a large wooden case, lined
inside and out with tinfoil, the metal surface reflecting radiation falling on it from outside and

radiating little to the inside, and so lengthening out and reducing fluctuations of temperature. The apparatus was in a closed cellar and the tilt of the balance beam was observed by a telescope through a hole in the floor of the room above.

The pans of the balance were removed and in their place two lead spheres A and B, each 6 inches in diameter and weighing about 21·6 kgm. or 48 lbs., were hung from the ends of the beam. These were the attracted masses. The beam was not lifted up from its support between weighings, as in the usual operations with a balance, but was left free to swing through a whole series of experiments, often extending over a number of days.

Underneath the balance was the attracting mass M, 1 foot in diameter and weighing 153·4 kgm. or 340 lbs. This was placed on a turn-table which could be rotated about an axis exactly under the centre of the balance by a rope passing to the observer in the room above. M could be brought against a stop so as to be exactly under A, with a distance of 1 foot from centre to centre, or it could be moved round against another stop so as to be exactly under B.

The attraction of M on A in the first position made A slightly heavier. When it was moved round to the second position under B, its attraction was taken from A and added to the weight of B, and the balance tilted over on the *B* side through an exceedingly small angle due to a change in the weight of *B* amounting to twice the attraction to be measured. It was necessary, then, to measure the small tilt and to find the change in weight, or the attractive pull, to which it corresponded.

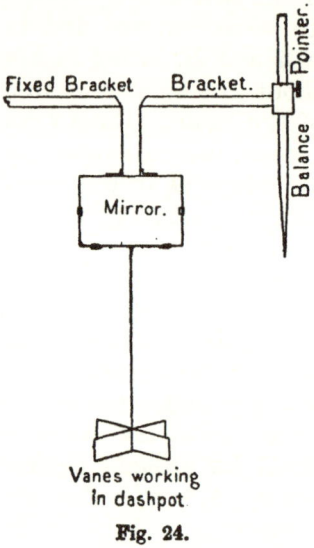

Fig. 24.

Firstly, to measure the tilt a 'double-suspension' mirror was used, a device due to Lord Kelvin. This was applied as shown in fig. 24. The beam of the balance must be supposed to be perpendicular to the plane of the figure some 2 feet above the end of the pointer. Near the end of the pointer a bracket was attached to it, and opposite to its mean position was a fixed bracket. A mirror was hung from the ends of these brackets by two silk threads. Now imagine that the balance beam tilts, say the further end downwards. The pointer will move out of the plane of the paper towards us and the mirror will turn round, and the angle through which it turns will be as many times greater than the angle through which the beam turns as the length of the pointer is greater than the distance between the suspending threads. The length of the pointer was about 150 times this distance, and the advantage of such magnification is obvious, as the tilt of the beam was not much more than a second of arc. To prevent the swinging of the mirror independently of the balance a set of vanes was attached to it below, working in a dashpot containing mineral oil. An inclined mirror, not shown in the figure, was fixed just in front of the suspended mirror. An illuminated scale was fixed to the telescope, and the light from this was reflected first from the inclined mirror to the suspended one, then back to the inclined one, and so up into the telescope. The observer saw the image of the scale moving up and down as the balance moved and noted the division on a cross-hair which was fixed in the eyepiece in the middle of the field of view.

Secondly, to determine the weight-value of the tilt two riders were used, each 1 centigramme in weight. One of these was lifted off the beam and the other was put on to it exactly 1 inch further from the centre, equivalent to a transfer of the first rider through 1 inch. As the half length of the beam was 24 inches, this was equivalent to an addition of $\frac{1}{24}$ of 10 mgm., about 0·42 mgm., at the end of the beam. The change in the scale-reading due to the change of rider was noted. It happened that it was very nearly equal to the effect of moving M from its position under A to its position under B, so that the pull of M on A at 1 foot was about 0·21 mgm. It is not necessary to describe the method of putting the riders on and off, but it may be mentioned that in order to secure a transfer of exactly 1 inch two little frames equivalent to two little scale pans hung from points equivalent to knife edges exactly 1 inch apart along the beam.

We shall not give particulars of the weighings. It will suffice to say that observation of the change in scale-reading due to shift of the rider was alternated with that due to change of position of M for a considerable number of determinations of each, and the means were taken. The mass M not only attracted the hanging masses A and B, but also their suspending rods and the arms of the balance. To get rid of these effects, a second set of measurements was made, in which A and B were put higher up the suspending rods in the positions A' and B', so that there was double the distance, viz. 2 feet, from centre to centre. The attractions on A and B were reduced to $\frac{1}{4}$ the previous amount, but the attractions on beam and rods remained as before. The difference between the two values was thus $\frac{3}{4}$ the value of the attraction in the lower position.

Of course there were cross attractions of M on B when it was under A and of M on A when it was under B, tending to reduce the effect observed. But the reduction could be calculated and allowed for.

Originally the mass M alone was on the turn-table, but some curious inconsistencies appeared in a series of results obtained, and ultimately it was found that these inconsistencies were due to a tilting of the cellar floor when the mass M was moved from one side to the other. The floor probably tilted through an angle about a third of a second, which would amount to 1 inch in 10 miles, and this tilt was quite enough to affect the results very seriously, as the whole tilt of the beam due to the change in attraction when M was moved round did not amount to as much as 2 seconds. The floor-tilt had been looked for before exact measurements were begun, but it had not been detected. It asserted its exist-ence later, and in such a way as to spoil a long series of measurements. It may be noted that if the tilt had always been the same, it would have been eliminated by the differential method of taking the attraction. But it grew as time went on, for the floor gradually settled down and became more com-pact, all tilting over together.

In order to prevent any tilt a second mass m (fig. 23) was introduced, having half the weight of M

and being at double the distance from the centre on the opposite side. The centre of gravity of the two was thus at the centre and the prevention of tilt of floor was complete. The attraction of m on A and B had now to be allowed for, but that only made the calculation of the results a little more complicated.

A very rough value of the mass of the Earth may be obtained thus: M attracted A, at an effective distance of 1 foot, with a force of 0·21 mgm. weight. The Earth attracted A at an effective distance equal to its radius of 21 million feet, with a force equal to the weight of A, i.e. equal to 21 kgm. or 10^8 times as much. Had the Earth been 1 foot away its mass would have been $10^8 \times$ mass of M or $10^8 \times 340$ lbs. But as it was 21 million feet away its mass was $(21 \times 10^6)^2$ times this or about 15×10^{24} lbs.—an over estimate due to inexact numbers and neglect of corrections. As the mass of an equal sphere of water is about $2·5 \times 10^{24}$ lbs. the mean density of the Earth is roughly 6.

The result obtained for the mean density after all corrections was 5·49.

It may be interesting to state the accuracy with which the balance worked. The increase in the weight of the 50 lbs. which was to be measured was about $\frac{1}{50,000,000}$ of the whole weight. Measurements of this increase were never wrong by more than 2 per cent. of the amount, usually well within 1 per cent., or $\frac{1}{5,000,000,000}$ of the whole weight, the variation which would occur if the 50 lbs. were moved $\frac{1}{40}$ inch nearer to the centre of the Earth. Now these numbers in the denominator are too large to give us much idea of the smallness of the weights concerned. Suppose, then, we take a rough illustration, in which the small weights are magnified up to be appreciable.

Imagine a balance large enough to contain on one pan the whole population of the British Islands, and that all the population has been placed there but one medium-sized boy. Then the increase in weight which had to be measured was equivalent to measuring the increase due to putting that boy on with the rest. The accuracy of measurement was equivalent to observing from the increase in weight whether or no he had taken off one of his boots before stepping on to the pan.

One of the most curious points about this method of weighing the Earth is the contrast between the mass to be weighed and the mass in terms of which it is weighed. It will be remembered that the tilt of the balance was measured by moving a centigramme rider along the beam. Any inaccuracy in the estimation of the weight of that rider is repeated in the weight of the Earth. So that in one sense we may be said to weigh the Earth with its 13 billion billion pounds by using a weight of $\frac{1}{50,000}$ part of a pound.

The results of all recent experiments, whether by the torsion balance or by the common balance, agree in giving to the mean density of the Earth a value very near to 5·5, and probably the real value is a little greater than this, but not so much as 5·55.

Though all the experiments have been described as if they were designed to find the mean density of the Earth, they have a more general aspect and may be regarded as determining the exact expression of Newton's Law of Gravitation. That law states that the attraction between mass M_1 and mass M_2 a distance d apart is proportional to $\dfrac{M_1 \times M_2}{d^2}$. Let the masses be measured in grammes, the distance in centimetres and the attraction P of either on the other in dynes. We may put the law in exact form as

$$P = \frac{G \times M_1 \times M_2}{d^2},$$

where G is a constant, the same for all masses, whether they be the Sun and Earth or the Earth and Schiehallion or the attracting and attracted spheres in any of the Cavendish class of experiments. It is called the Gravitation Constant.

Now as any of the Cavendish experiments consists in determining P between known masses M_1 and M_2, d apart, the result gives us G at once. It may be shown that the other Earth-weighing experiments also give G, though not quite so directly. The value of G is very near to $\dfrac{6\cdot66}{10^8}$.

CHAPTER III

THE EARTH AS A CLOCK

EVERY day the telegraph lines over the whole country cease work for a short time for the passage of a signal which is sent out from the Observatory at Greenwich exactly at 9 a.m. At the Observatory there is a Standard Clock, and that Standard Clock is the Earth itself. The sky is the dial. Its figures are the stars, and the line of sight of a telescope is the hand which points the hours. What sort of time does this clock keep?

If we face southwards on a clear day we note that the Sun has risen on our left, mounts to the highest point in the south, and sinks down to set on our right. On a clear night we note that if we still face southwards the Moon and stars move in the same way from left to right. We feel at rest, and we see the lights of day and night moving over us.

Till 300 or 400 years ago almost everyone believed that this was the only and final account of the appearance, that the Earth was at rest and that the sky moved round. But now we are certain that a more convenient and therefore a better account is that the appearance is due to the Earth turning round under the sky. It is no wonder that this new account had a hard fight against the old belief, and that it only slowly conquered. It is impossible to realise that we are being whirled round in a huge circle, travelling in this latitude of Britain at a speed of 600 miles an hour. It seems at first thought as if we should be whirled off into space. But it is easy to show that a very minute fraction of our weight is sufficient to keep us from so flying off.

The feeling that we are at rest while the sky is moving over us is just like that which we have when we are seated in a smoothly running train and see the buildings and telegraph posts rushing past us; or when we are in the cabin of a steamer on a river and, looking out of a porthole, see the river banks drifting past us. If our only aim is to describe what we see in change of relative position, it is perfectly correct to say that we are sitting still in the train or boat and that the country is moving past us. And in a similar way it is perfectly correct to say, if we are only describing the relative change of position, that the Sun, Moon and Stars rise in the east, climb up the sky and set in the west.

But when we come to consider not only change of relative position but the forces which effect the change, then we are obliged to think of one description as better than the other. Our train stops at a station, and we can think of the friction of the rails against the braked wheels as stopping it. We cannot think of a force which would stop the station at the train. Our steamer stops at a pier, and we can think of the pulls of the mooring ropes acting to stop it. It is impossible for us to think that the whole countryside is pulled up alongside the steamer. Similarly with the Sun, Moon and stars. We know that the force needed to keep a body moving in a circle is proportional to its distance and to the square of its rate of revolving round the centre. We can easily think of the weight of bodies as sufficing to keep them on the surface of the Earth if it whirls round once in 24 hours. But if the Earth were at rest and the heavenly bodies were moving round it once in 24 hours, it would indeed be difficult to imagine forces big enough. Each pound of matter in the Moon, to get round a circle 240,000 miles in radius in 24 hours, would require a pull of about $\frac{1}{3}$ lb. weight. Each pound in the Sun, to get round in its circle, would require a pull of about 80 lbs. weight ; each pound in Sirius, to get round in its gigantic circle, from which light only comes to us in 8½ years, would require a pull of something like 20,000 tons. We cannot contemplate the possibility of these huge forces: so that unless the Sun, Moon and stars are mere phantasms and not real matter, we are obliged to think that the motion is in the Earth and that it is whirling round under the sky.

We have direct evidence of this whirling. The shape of the Earth as described in the first chapter is a consequence. Suppose that a perfect sphere the size of the Earth were suddenly started spinning once in 24 hours round an axis. Part of the weight of the matter in the equatorial regions would be used up, as it were, in keeping it moving in its circle. It would press less towards the centre. The matter at the poles, not moving round, would press with its whole weight, with the result that it would press out the equatorial matter and make it bulge, and an equatorial bulge is just what the shape of the Earth shows.

Another consequence is the direction of spin in cyclones, the vast whirlwinds which are such common features of the weather in this part of the globe. The centre of a cyclone is a point where the barometric pressure is lower than anywhere in the neighbourhood and the wind circles round the centre always in the same direction, counter-clockwise as plotted on a map in the northern hemisphere, clockwise in the southern hemisphere.

Let C (fig. 25) be the centre of such a cyclone, the point of lowest pressure. Let the circle $NWSE$ be a line on the map passing through points round C at which the pressure has a certain higher value. At first thought we might expect the air to be pressed straight in towards C from all sides. But the wind does not blow straight in to the centre

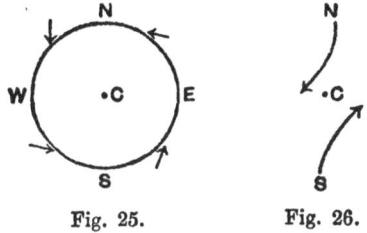

Fig. 25.   Fig. 26.

of low pressure. It is observed much more nearly to circle round it. In general it is inclined somewhat inwards, as indicated by the arrows in fig. 25, and always in our latitude the whirling is counterclockwise.

The way in which cyclones are formed is not yet understood, but the following explanation of the direction of the whirling is probably correct. Consider a mass of air at S (fig. 26) to the south, which tends to move to C as it moves northwards.

It keeps moving into regions travelling less rapidly to the east than the regions from which it has come. It keeps some of the excess of its W. to E. motion, and so instead of moving due north to C it moves partly to the E. or, on the whole, to the N.E., and so may have the direction of the lower arrow, fig. 26.

Next consider the motion of a mass of air from N. As it moves southwards it will be continually moving into regions with a greater W. to E. motion than its own. It will lag behind and come partly from the E. as well as the N., and so may have the direction of the upper arrow (fig. 26).

Thus there is imparted a tendency to whirl round the centre and always in the same direction; the winds tending to go to the right of the centre, and so starting a counter-clockwise rotation to the cyclone.

If the case of the Southern hemisphere be considered it will easily be seen how it comes about that the whirling of the cyclones there is in the opposite or clockwise direction.

The shape of the Earth and the whirling of cyclones constitute observational evidence for the rotation of the Earth. We owe to Foucault[1] two experimental methods of proving the rotation. The first of these is that of the Foucault pendulum. To understand it let us assume that the Earth is

[1] *Recueil des Travaux Scientifiques de Léon Foucault.*

spinning round the polar axis POP (fig. 27). Take any point A on the surface and draw the two lines OA and OB through the centre O at right angles to each other, and in a plane through POP; we may resolve the spin about OP into two spins going on at the same time about OA and OB.

It is important to note that the spin of a body *round* an axis can be represented by a length *along*

that axis proportional to the number of turns in a given time. The rule for resolving a spin is as follows. If OP is taken to represent the rate of spin round PP, once in 24 hours, dropping perpendiculars PM on OA and PN on OB, the spin round OA is represented by OM, while that round OB is represented by

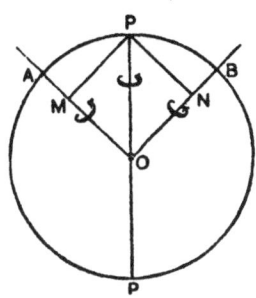

Fig. 27.

ON. In general OM and ON are less than OP, so that the spins which they represent are slower than once in 24 hours. Thus the surface of the Earth at A is being carried forward, out of the plane of the figure, by the motion round the axis OB, while at the same time it is turning round the vertical OA with a spin OM. When A is at the pole this turning around the vertical is a maximum, for there $OM = OP$, and a revolution is effected in

24 hours.   As we move A down towards the equator, OM decreases.   In our latitude the ground turns round the vertical once in about 31 hours.   When A is at the equator, OM vanishes and there is no turning round the vertical.

Foucault's experiment consists in hanging up a heavy pendulum by a wire many feet long and setting it swinging in a definite vertical plane.   As the surface of the ground and, with it, the support of the pendulum turn round the vertical, the only action on the pendulum is slowly to twist the wire, and this merely twists the bob round its vertical axis.   It goes on swinging in the plane in which it was started, and the ground revolves underneath it in a counter-clockwise direction.

But to the observer moving round with the ground the plane in which the pendulum bob swings appears to move round clockwise as looked at from above.   With a pendulum some 30 feet long, swinging across a horizontal circle of 4 feet radius drawn round its lowest point as centre, the bob moves, each time of return, to a point about $\frac{1}{80}$ inch further round the circle so that the motion is evident in quite a few swings.

Foucault's second mode of showing the rotation of the Earth was by means of the gyroscope, a heavy disc which can be set spinning on specially arranged supports.

When a body is acted on by a force through its centre of gravity the force does not tend to turn the body round, but merely to move it forward. If the force is in a line passing to one side of the centre of gravity, a 'sideway' force as we may term it, then it tends to spin the body round as well as to move it on. But when the body is already spinning round an axis the effect of the sideway force in changing the direction of that axis is less the greater the spin. For instance, let a force act to one side of the centre of gravity of a sphere, which in one second will give

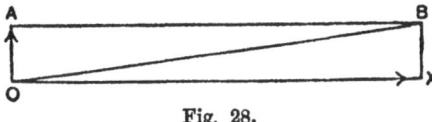

Fig. 28.

a spin round OA, fig. 28, represented by OA. If the body was previously at rest, a line OX will revolve round OA at a rate proportional to OA. But now let the body be revolving round OX with a spin represented by OX. Then we must compound the two spins OA and OX, and their resultant is OB, or the new spin is about OB, and at a rate represented by OB. The greater OX is in comparison with OA the nearer OB is in magnitude and direction to OX.

The gyroscope which Foucault used was arranged as represented diagramatically in fig. 29. D was a

heavy disc whose axis was pivoted on a circle *CC*.
This circle was supported in turn by knife edges *KK*,
on which it was exactly balanced, and these rested
in V-shaped hollows on a second circle *EE*. This

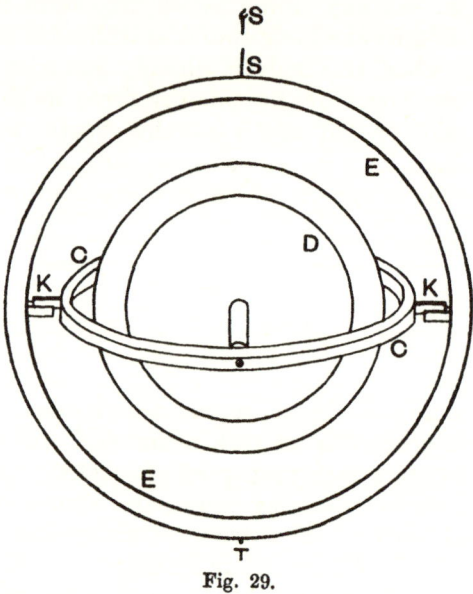

Fig. 29.

circle was suspended by a silk fibre *SS*. The disc,
then, could be turned into any position subject to the
limitation that the knife edges could not allow a
very great tilt; for it could be rotated about either

of three axes at right angles. The disc was set spinning within the circle CC with very great rapidity, and was then put in position with KK resting on the V's. The pull of the string up and the weight acting down both passed through the centre of gravity and did not tend to give any spin to alter the direction of the axis of rotation of the disc. Any friction which might come into play as a sideway force would be so small that it would be very slow in altering the direction of the axis since the rate of spin was very great. The direction of the axis of spin of the disc therefore tended to persist in the same direction in space. If, for instance, it was pointed at a particular star, it remained pointing at that star. Or let us suppose that the axis pointed at any star on the horizon. As a star begins to rise, in general it moves partly up and partly along the horizon, and the latter component of the motion can be shown to be the same for all stars just rising to an observer at the same place. Foucault directed a microscope to a finely divided scale placed across the outer circle near K, and he found that if the disc was set spinning and had its axis directed to any point on the horizon it turned round always in the direction N.E.S.W., and its motion along the horizon was the same as that of a star rising at the point; the rotation having the same rate, in fact, as that of the plane of his pendulum.

Foucault also considered what should happen to a gyroscope in which the axis of spin can only move round horizontally.   We may think, for example, of the gyroscope in fig. 29 as having the inner circle CC fixed to the outer circle EE so that the axis of the disc D is horizontal. As the Earth rotates, the vertical is continually changing its direction in space. If the gyroscope were not spinning, its centre of gravity would be pulled at once by the weight into every new position of the vertical through the point of suspension; we should have merely a plumb-bob and line.   But when the disc is spinning very rapidly, the tendency of the axis to persist in its direction does not allow this immediate adjustment to take place, and the centre of gravity is in general not directly under the point of suspension. Thus the weight and the pull of the string not being exactly in one line but in parallel lines, impart a little spin which, compounded with the spin of the disc, tends to make the axis move into the N. and S. line. A special case must serve as illustration, for the complete action is hardly to be followed without mathematical representation.   Let the gyroscope be on the equator, and let the axis of its disc be horizontal and directed east and west. Fig. 30 (a) is an elevation as seen from the south, VV being the vertical.   Let us suppose that the top part of the disc is moving southwards, and that the spin is represented by OX,

fig. 28. Now let the instrument be carried round by
the Earth's rotation to a region fig. 30 (*b*) where the
direction of the vertical is $V'V'$. If the direction of
the axis of spin persisted, the pull of the string and
the weight would have to be as in fig. 30 (*b*), and
they would give a spin round the N. and S. line in
which the top part of the disc would move eastward.
The tendency to persistence in direction of spin

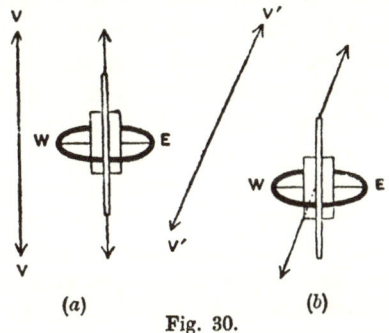

(*a*)   (*b*)

Fig. 30.

introduces, then, a new spin about an axis at right
angles to the initial spin, and if we represent this
new spin by OA, fig. 28, the resultant is OB, *which
is nearer to the meridian.* It can be shown that in
whatever direction we have the axis to start with, the
action is of the same kind, tending to bring it nearer
to the meridian. When the meridian is reached,
the motion gained carries the axis through it and,

were there no friction, the axis would go as far on the other side, would then return and would continue to vibrate to and fro. But through friction the vibrations will gradually be lessened and if the spin of the disc is maintained it will ultimately settle down pointing true north. Foucault could not maintain the spin of his disc sufficiently to verify this, but he expounded the principle very clearly.

It has lately been carried out in practice in that very remarkable invention by the brothers Anschütz, termed the Gyro Compass[1]. In this compass the gyroscope disc is represented by a 3-phase electric motor, to which the current is fed through the suspension, and the spin of the motor is maintained at about 300 revolutions per second There is a special arrangement for damping out the vibrations about the N. and S. line, and the axis of spin ultimately settles in that line. The instrument thus constitutes a mariner's compass, pointing always to the true north. It is, of course, quite free from the deviations due to iron or steel which the magnetic needle displays. It appears to act wonderfully well, and if the expense of its construction could be lessened it would no doubt entirely displace the magnetic compass on all large ships.

Both observation and experiment thus confirm the supposition, already inevitable from a consideration

---

[1] *The Anschütz Gyro Compass.* Eliott Brothers, London, 1910.

of forces, that the Earth spins round its axis. Observation further shows that it spins at a rate so nearly approaching uniformity that the time of one revolution does not change by more than a quite immeasurably small fraction of a second in a year.

We can see how the uniformity is preserved by considering how spins are made or are changed. They may be made or changed by the action of forces not passing through the centre of gravity of the body spinning, or they may be changed in rate by alteration of shape of the body.

To illustrate the mode of making or changing spins by sideway forces, let us suppose that we hang up a ball by a string and deliver a blow full on it directed through its centre; it merely vibrates to and fro, pendulum-wise. But if a peg projects from one side and we deliver the blow sideways on the peg, the ball moves pendulum-wise and at the same time spins. Thus a spin is made by a sideway force, a force acting not through but to one side of the centre of gravity. If while the ball is spinning we hit it full we may so time the blow and its strength that we stop the swing, but the spin still persists, for the force applied was through the centre of gravity. But if we hit the peg we give a blow at one side of the centre of gravity. We shall then alter the spin and may even stop it.

Now the Earth is acted on by forces from outside,

and chiefly from the Sun and Moon. The resultant attractions of these bodies go almost exactly through its centre—and at present we will suppose they go quite exactly through it—and so they are unable to alter the spin.

We shall see later that we are obliged to suppose

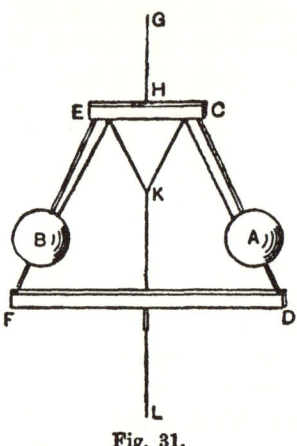

Fig. 31.

that this is not quite true, though we have no certain evidence as yet of alteration of spin.

Again, if a spinning body changes its shape its spin may change, and so if the Earth were, for example, decreasing its equatorial bulge, its rotation might be speeding up. Any bringing together of the matter of a spinning system towards the axis of rotation makes it go round the axis in a less time. We may illustrate this by the apparatus represented in fig. 31, where A and B are two balls sliding on rods CD and EF, the two sides of a frame $CDFE$ hung by a string GH. To A and B are attached strings coming together at K, and to K is attached the string KL. Now let the frame and balls be set whirling round the axis *GHKL*. If the string *KL* is pulled down, *A* and *B* move nearer to the axis and the balls move round more turns per second. If *KL* is released *A* and *B* move down and further from the axis, and the balls move round fewer turns per second.

With the Earth just the same principle holds. If it were contracting at a sensible rate the spin would increase and the day would decrease. It is very probable that the Earth has contracted in the past, and it may be very slowly contracting now. But if so, the rate is so slow that any quickening which it might have produced in the rotation is probably more than counterbalanced by a slowing down which the tides have produced in a way explained hereafter.

In any case the effect is very minute, as certain ancient records of eclipses show. Eclipses occur in series at definite intervals, which we can express in terms of the present time of rotation of the Earth, and so we can reckon back from eclipses observed at the present time to eclipses which ought to have occurred in the past. Some few actual eclipses recorded by Assyrians, Babylonians, Egyptians, or Greeks, appear to agree very nearly with such calculated eclipses. Thus there is one series of eclipses of the Sun at intervals of very nearly 29 years, and the sum of 18 of these intervals is almost exactly 521 years. There was one of this 521 year series in 1843.

Reckoning back, a total eclipse should have occurred on June 14, B.C. 763, and the path of totality should have passed 100 miles or more north of Nineveh. We have a record of a total eclipse at Nineveh about this time, in all probability that calculated. There are, however, certain difficulties in making the calculations exactly fit the records of these ancient eclipses. Thus observation made totality in B.C. 763 at Nineveh; calculation makes it to the north. But Mr Cowell has shown that calculations fit the records much better if the day is lengthening by $\frac{1}{200}$ second per century! If we assume that this is occurring, going back a century, the average day during the century is $\frac{1}{400}$ second shorter than the present day, and as there are 36,500 days in the century, the actual century will be shorter than a century made up of our present days by about $36500/400 = 90$ seconds. Going back 2500 years, nearly arriving at the Assyrian eclipse, 25 centuries of our days would exceed the actual 25 centuries by $365 \times 2500 \times 25/400$ seconds or 15 hours. There is no reason to suppose that the change has been much greater than this. So that it is probably safe to conclude that the Earth is rotating so uniformly that it has not lost nearly so much as a day in 2500 years.

The Earth, then, spins round practically uniformly under the sky, and the fixed stars appear in consequence to return to the same place at equal intervals. But our ordinary day is not exactly one of these intervals. It is fixed by the interval between successive returns of the Sun. When, however, we time the Sun's return to the meridian, to the south line in the sky, by a very good clock we find that it does not give us uniform days. Our actual 24 hours' day is only the average interval between successive passages of the Sun across the meridian.

Let us suppose that we have a uniform clock, a 'Mean Solar Clock,' of which 24 hours is exactly the average interval between the Sun's successive passages across the meridian or south line throughout the year. Then according to the clock the Sun is sometimes fast, sometimes slow, and for two reasons which we can examine separately.

The first reason is that the Earth moves round the Sun in an ellipse, with its greatest speed when nearest the Sun and its least when farthest away; and the second reason is that the Earth's axis is not perpendicular to the plane of its orbit.

Taking the first reason, let fig. 32 represent the orbit of the Earth, its ellipticity being grossly exaggerated, and, as we are treating the two reasons separately, let us suppose that the Earth is spinning round an axis perpendicular to the plane of the orbit. Let A be the Earth's position when nearest the Sun, when it is moving fastest, and let P be the point where the Sun is due south. Next day when the

Sun is due south for the same point P let the Earth
have moved to B. We enormously exaggerate the
distance AB in the figure. Then we see that the
Earth has moved more than once round, and by
the angle SBE, where BE is parallel to SA, or by
the angle BSA. Now let us go round to the other
side of the orbit six months later when the Earth

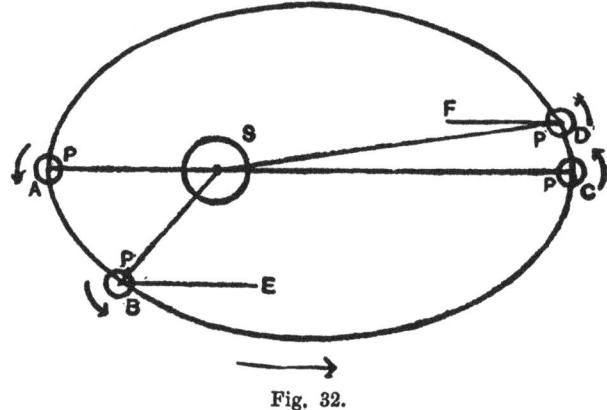

Fig. 32.

is at C and the Sun is due south for the point P.
Next day when it is due south for that point let the
Earth have travelled CD, which is less than AB.
The Earth must have spun more than once round
by the angle SDF, where DF is parallel to SC,
or by the angle DSC, which is obviously less than
BSA. Hence the time between two successive

southings of the Sun is less when we are at C, that is about July 1, than when we are at A, that is about December 31.

Now let us consider the second reason for the unequal lengths of the solar day, viz. that the Earth's axis of spin is not perpendicular to the plane of the Earth's orbit round the Sun. Let us assume that the last investigation has shown us the effect of varying speed in an elliptic orbit, and that we may investigate the present effect while supposing that the Earth goes round the Sun at uniform speed in a circle. We can see at once that if the axis were perpendicular to the plane of the orbit then the times between successive southings of the Sun would always be exactly equal for a given point on the Earth's surface, and for every point day and night would each be 12 hours. As the Earth went round the Sun, the Sun would appear to go round the Earth in the equator of the sky, half way between the sky poles. But the Earth's axis points about 23° away from the perpendicular to the plane of the Earth's orbit or the plane of the ecliptic. The Sun, therefore, in its yearly course round the sky does not appear to move round the equator but in the circle in which the plane of the orbit meets the sky.

Let the Earth spin counter-clockwise round its axis PP (fig. 33) at the centre of the equatorial sky

circle *EBQA*, and let *ACBD* be the circle in which
the plane of the orbit cuts the sky—the ecliptic.

The relative positions of Earth and Sun will be
just the same if we keep the Earth at the centre and
make the Sun move uniformly and counter-clockwise
round the circle *ACBD* or the ecliptic. Twice in

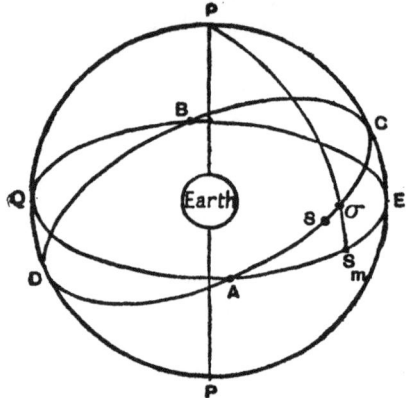

Fig. 33.

the year—on March 21 at *A* and on September 23
at *B*—the Sun is on the equator, and day and night
are equal all over the Earth.

Now let us set an imaginary sun—we will call
it the mean Sun—to travel at uniform speed round
the sky equator *AEBQ* and so that it goes through
A and again through *B*, with the real Sun moving

round the ecliptic $ACBD$. Then if this imaginary
sun could replace the real Sun evidently day
and night would always be equal, and if we constructed a perfect clock to show 12 noon for two
consecutive passages across the meridian of the
mean Sun it would show 12 noon for every passage.
Such a clock would be said to keep Mean Solar
Time.

If $P\sigma S_m$ is the meridian of a certain place on
the Earth's surface, this meridian must be supposed
to sweep round the sky from left to right as seen
from the outside. When it meets the mean Sun S_m
then for that place the time is mean noon.

Now starting the mean Sun S_m and the true Sun
S from A on March 21, they move from left to right
as seen from outside, and a short time later when
one is at S_m the other is at S, where $AS = AS_m$. It
is evident from the figure that for a short time AS
is less than $A\sigma$. This means that the meridian as it
moves from left to right as looked at in the figure
meets S before it meets S_m. Then the true Sun is
before the clock. But by June 21, when the mean
Sun is at E the true Sun is at C, and the meridian
again meets them at the same instant: thus between
March and June there is a time, which is early in
May, when the true Sun is a maximum amount in
advance of the clock. As we have seen, on September 23 the two suns coincide, but a little time before

that it is easy to see that the meridian meets the
mean Sun first or the true Sun is behind the clock.
Similarly it can be seen that between September and
January it is in front, and between January and
March it is behind.

Thus we have two effects, that due to ellipticity,

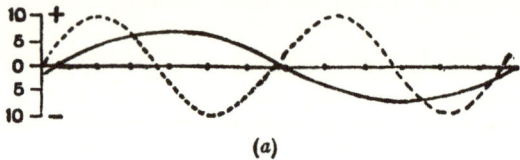

(a)

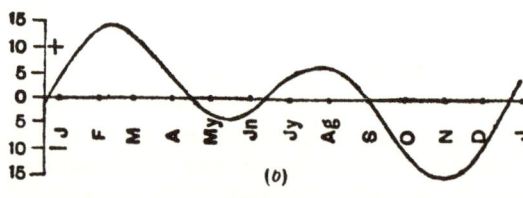

(b)

Fig. 34.

which makes the Sun before the clock half the year
and behind the other half, and that due to inclination
of the axis, which makes the Sun before in one
quarter and behind in the next. In fig. 34 a we represent these two effects separately, the continuous
line being the ellipticity effect, and the dotted line
the inclination effect. In fig. 34 b we represent the

sum of the effects.   These figures are taken from
Godfray's *Astronomy*.

The nett amount by which the true Sun is behind
or in front of the mean Sun or the perfect clock
is called the 'equation of time,' and it is reckoned
positive when the Sun is slow, negative when the
Sun is fast.

A sundial keeps true solar time, and so the
'equation of time' has to be added to its indication
to give the clock time.   When the equation of time
is positive the sundial is slow; also when it is positive
sunrise is nearer clock noon than is sunset, and this
is very noticeable in January.

The Sun, then, though in the long run he rules
the length of the day, does not keep regular time
day by day as tested by a uniform clock, and so he
fails us as a regulator.   How are we to test whether
our clock is uniform ?

For this purpose we must use the fixed stars,
which come round night after night to the south
meridian at very nearly equal intervals.   They are
watched and the clocks are rated by them at
Greenwich and at other observatories.   But the
hand of the sky clock is not the line to any one
star, nor even to a point fixed relatively to the
stars.   It is the line to the point A in fig. 33, the
'first point of Aries,' and this point travels slowly
round the equator, completing its circle in 25,800

years. This is due to the fact that the Earth's axis is not in a fixed direction but is moving round the perpendicular to the orbit, and A in fig. 33 is moving round $DACB$. The Sun and Moon act on the equatorial bulge of the Earth in such a way as to make the Earth wobble or precess, and the wobble, accomplished in 25,800 years, is superposed on the spin. Through it the whole sky slowly rolls round the perpendicular to the orbit, and that is why any particular star is unsuitable for a perfect timekeeper. The time between the successive passages of the first point of Aries across the meridian is 23 hours 56 minutes 4·09 seconds, and a 'sidereal' clock which is perfectly rated by the stars should show 24 o'clock every time the first point is due south.

To get the time of passage of the first point, an ever shifting point, from observations on so-called fixed stars, calculations are necessary, depending on the position of the star observed. In practice the calculations are turned the other way about, so as to give the times of passage of certain stars across the meridian after the passage of the first point of Aries. These times are set forth in the *Nautical Almanac*, and the stars are called 'Clock Stars.'

In making the determinations two other disturbing effects must be taken into account. Over and above the effect in precession the moon produces another very slight wobble in the direction of the Earth's axis, which is accomplished in 19 years so that not only does the sky roll slowly round but it quivers as it rolls. This 19 years' quiver is called 'nutation.'

The other effect is due to the finite speed of travel of light and it is termed 'aberration.'

For our present purpose we may describe aberration as the alteration in the direction in which something appears to come to us as we change the direction or the speed of our own motion.

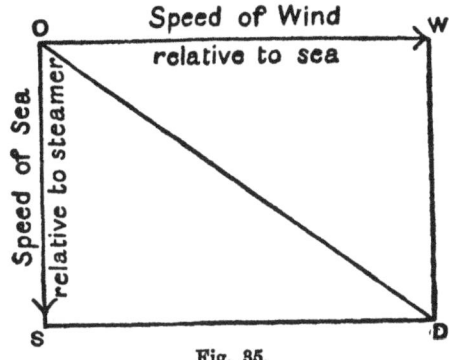

Fig. 35.

We have an excellent example of aberration in the direction of the wind on a steamer. Suppose that for a man on a steamer at rest the wind is from the west. If the steamer is travelling due north, then to an observer on the steamer the wind will appear to be coming from somewhere between N. and W. The motion of the air relative to the steamer may be obtained by drawing from a point O (fig. 35) one line OW representing the velocity of the wind relative to the sea, and another line OS representing the reverse of the velocity of the steamer, really the velocity of the sea relative to the steamer. Drawing a parallelogram on these two lines its diagonal OD represents the direction of the wind relative to the

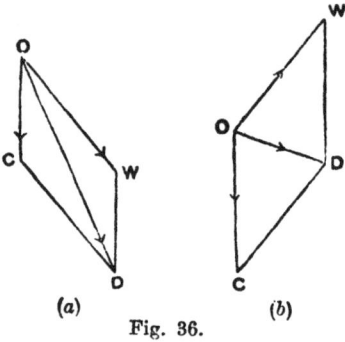

(a)   (b)

Fig. 36.

observer on the steamer. The change in the direction of the wind from OW to OD is its aberration.

A similar aberration of wind is a common experience of every cyclist. His motion turns a wind coming from any point in the front into one more nearly a head wind and a wind from behind is lessened in its effective speed, or may even be turned into a head wind.

In fig. 36 a or b let OW represent the velocity

and direction of the wind over the surface of the
ground, *OC* the velocity of the ground towards the
cyclist, that is the reverse of the velocity of the
cyclist over the ground ; then *OD*, the diagonal of
the parallelogram on *OW* and *OC* represents the
wind as experienced by the cyclist. In fig. 36 *a* we
see how the wind is made much more of a head
wind and much stronger. In fig. 36 *b* a wind partly
from behind becomes one with a small component
from the front.

There is a precisely similar effect, however it may
be produced, in the case of light. Just as in the
cases we have considered, our motion as we are
carried round the Sun alters the apparent direction
of the light which we receive from any star. We
have to compound with the velocity of the light the
velocity of the earth *reversed* and the resultant of
these gives us the direction in which the light appears to come to us.

Let *ABCD* (fig. 37) represent the orbit of the
Earth as seen from the north side, the Earth going
round counterclockwise. Consider the light coming
from a star a very long way to the right in the line
CA.

Let the velocity of light be *V* and that of the
Earth be *v*.

Drawing parallelograms at *A* and *C* with sides *V*
and reversed *v*, we get the diagonals *SO* and *S'O'*

as the directions in which the star is seen. At *B*, *V* and − *v* are opposed, but the direction is not altered, while at *D* they are added to each other, and again without alteration of direction. At an intermediate point such as *K*, the effect is intermediate between that at *C* and *D*. Thus the star appears to shift to

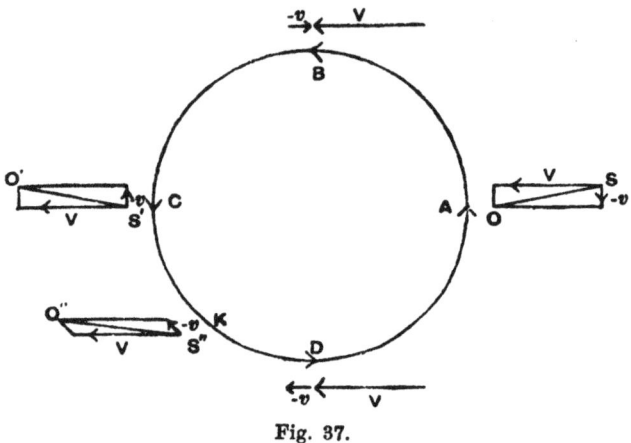

Fig. 37.

and fro in the course of the year. The speed of the Earth *v* is only about $\frac{1}{10000}$ the speed of light *V*, so that in fig. 37 the change of direction is enormously exaggerated. In fact the star as seen from *A* and from *C* will only be about 20 seconds of arc on either side of its position as seen from *B* or from *D*, the amount subtended by the swing of a pendulum which moves one foot to either side of its lowest point when seen from a distance of two miles. It will be seen from fig. 37 that in the six months when we are nearest to the star it is deviated to the left, or the east, and so passes the meridian a little late. During the other six months it is to the right or the west, and so it passes the meridian a little early. For stars in other parts of the sky the effect is a little less easy to explain and we shall be content to state that they appear to move in ellipses always with maximum excursion of 20 seconds on each side of the mean position.

The aberration is not only important as giving us a correction to the times of passage of stars across the meridian but also as giving us an excellent method of determining our distance from the Sun. It has been most carefully measured and its amount —the 20 seconds or thereabouts of its excursion— is known probably with great accuracy. This 20 seconds, as may be seen from fig. 37, is equal to v/V or the ratio of the velocity of the Earth round the Sun to the velocity of light. But experiments have been made which have determined the time taken by light to traverse measured distances on the Earth's surface, i.e. which have determined V. Hence aberration gives us the velocity v of the Earth in its orbit. We can therefore find the distance it travels in a year, or the length of its orbit, and thence the radius.

The sky clock, when we allow for the roll of precession, the quiver of nutation, and the still smaller quiver of aberration, keeps time so perfectly that no terrestrial clock can detect any variation in its rate. But we must not depend on any one particular star. For the stars are undoubtedly moving, relative to our solar system, with velocities usually of the order of 10 to 50 miles a second and, in a few cases, with far higher velocities. In the nearer stars the components of these velocities perpendicular to our line of sight produce displacements which become visible in the course of years, and a star will gradually gain or gradually lose on the perfectly rated star clock owing to its 'proper motion' across the sky. But even in the nearest star this effect is very minute in the course of a year and absolutely unmeasurable in a single day. In the more distant stars it will not be detected even in the course of ages.

Some of the figures in our dial then are changing their positions. But we can detect these changes in course of years by watching the changes of pattern relative to the background of far more distant stars, and allow for them.

We are justified, then, in concluding that, as far as any present records go, the Earth spins practically at a uniform rate beneath the sky.

Has the Earth been always spinning, and will it continue spinning, at the same rate?

This question admits of an answer definitely in the negative.

We are certain that there is a slowing down of the spin due to the tides raised in the ocean by the Moon and Sun, even though it has been so infinitesimal during any time in which we have records available to show it, that we cannot be sure that it has amounted to a measurable quantity. The investigation of this slowing down we owe chiefly to Sir George Darwin.

To understand how it occurs, we must examine how the tides are formed, and how they follow the Moon and Sun as gigantic waves round the Earth.

The combined action of Sun and Moon, the irregular configuration of the oceans, their varying depths, and the variations of the tidal effect with latitude, all conspire to make the actual tides exceedingly complicated. We shall therefore idealise, and selecting only the part of the tide due to the Moon we shall suppose that the Earth and Moon move round each other in the Earth's equatorial plane and that the ocean forms a continuous canal round the equator of uniform depth, not more, say, than three or four miles.

Though we usually speak of the Moon as going round the Earth, in reality the two form a doublet revolving about their common centre of gravity, each in its own circle, in a month of $27\frac{1}{3}$ days, and it is this common centre of gravity which pursues, as it were, a smooth elliptic orbit round the Sun. The Earth and Moon swing now to one side, now to the other, of the orbit, but they are, of course, always on opposite sides of the centre of gravity round which they swing. We shall leave out of account the forward motion in the orbit and merely consider that which alone concerns us here, the monthly revolutions round the common centre. As the Earth's mass is about 80 times that of the Moon, the common centre of gravity is about 80 times nearer to the centre of the Earth than it is to the centre of the Moon, i.e. the two distances are about 3000 miles and 240,000 miles. Thus the common centre of gravity is about 1000 miles within the surface of the Earth.

We can perhaps see how the mutual pulls of the Earth and Moon suffice to guide them in their respective circles by the following consideration. A proposition in Mechanics tells us that the motion of the centre of gravity of any system is the same as if its whole mass were collected there, and all the forces acting on the system were transferred there unchanged in magnitude and in direction. The centre of gravity of the Earth is at its centre, and, as Newton proved, the forces on it due to the Moon are equivalent to the single force which the Moon

collected at its centre would exert on the Earth
collected at its centre. So the centre of the Earth
moves as if the Earth's mass were all collected there
and pulled by the Moon with a force which is proportional to

$$\frac{\text{Earth's Mass} \times \text{Moon's Mass}}{\text{Square of distance between their centres}}.$$

Similarly the Moon's centre moves as if all its mass
were collected there and as if it were subjected to a
pull equal and opposite to the above.

As the mass of the Earth is 80 times that of the
Moon and as the pulls on the two are equal, each
pound of the Earth is subjected to $\frac{1}{80}$ of the pull to
which each pound of the Moon is subjected. The
guiding force per pound being $\frac{1}{80}$, the circle which
it describes will only have a radius $\frac{1}{80}$ of the radius
of the Moon's circle. That is, the equal and opposite
pulls just account for the two bodies going in circles
round the common centre of gravity somewhat as
represented in fig. 38, though that figure is not
drawn to true scale. EE' is the Earth; C is its
centre; CC' is the circle which the centre describes
in $27\frac{1}{3}$ days round the common centre of gravity G;
M is the Moon and MM' an arc of its circle.

The Moon-pull is not the same throughout the
Earth. It is only *on the average* the same as at the
centre: being greater on the near, and less on the

far parts. This inequality leads to a deformation of the Earth. Deformation occurs, no doubt, in the solid body, which is to some extent yielding and not perfectly rigid, but it is more conspicuous in the ocean, loose on its surface, and therefore more easily yielding than the solid earth.

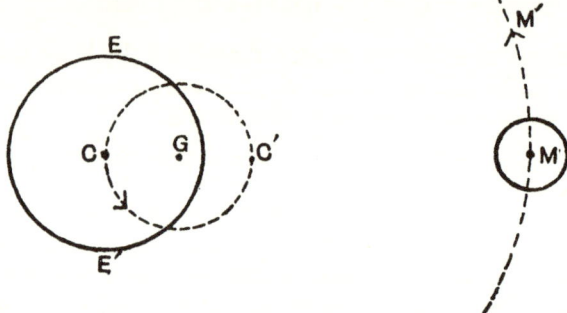

Fig. 38.

In considering the effect of the varying Moon-pull, let us suppose, in the first place, that the Earth presents always the same face to the Moon, just as the Moon presents always the same face to the Earth, so that, while it revolves round the common centre of gravity once in 27⅓ days it rotates also about its axis in the same time.

Let fig. 39 represent an equatorial section of the Earth, C its centre, G the common centre of gravity of Earth and Moon, and P any particle of matter in the equatorial canal.

The pull required to guide a body in a circle so that it shall get round in a certain fixed time is proportional to the radius of the circle. If, then, we represent the pull on each pound at C by the radius CG of its circle, the pull on each pound at P is represented by the radius PG of its circle. If we draw the parallelogram $CPMG$ the pull PG may be resolved into the two pulls PM and PC, and we can at once

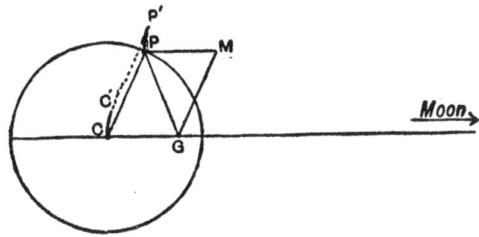

Fig. 39.

see the significance of each. The former $PM = CG$ is the same in magnitude and direction for every particle of given mass in the canal and is the force needed to guide it in a circle of radius PM equal to GC. But if we have revolution, without rotation, round G— the kind of motion exemplified by a bicycle pedal when the cyclist keeps his foot horizontal—every particle does go round in a circle, and the radius of every circle has the same value PM or CG. Thus while C (fig. 39) moves to C' round G, P moves to P' round M, where $C'P'$ is parallel to CP. So that the PM component is what is needed for revolution without rotation. The pull represented by PC is that needed to guide the matter at P in a circle with radius PC in a time of $27\frac{1}{3}$ days; that needed for the rotation apart from the revolution. It has nothing to do with the Moon. It is supplied by the Earth's attraction. Or in other words, some of the weight is used up in guiding the mass at P in its rotation circle. The consequence of this virtual reduction of weight is, as we have seen in Chapter I, a tendency to an equatorial bulge all round. If the Earth's speed of rotation is increased one effect is an increase in this component and an increased equatorial bulge.

As far then as the Moon's action goes, we need only regard the component $PM = CG$ and consider how far the Moon supplies this pull. In fig. 40, let $A'A$ be the equatorial diameter passing through the Moon and BB' the equatorial diameter at right angles. On the hemisphere facing the Moon its actual attraction on every particle is greater than it would be on the same particle at C except near B and B', i.e. it is greater than the attraction represented by CG. There is thus an excess over what is needed to keep the surface matter in its circle of radius PM, and the excess gradually increases from B where it is practically zero, to *A* where it is a maximum. Similarly it increases from *B'* to *A*. On the hemisphere away from the Moon the attraction is less than at *C*, so that we may represent it by a pull

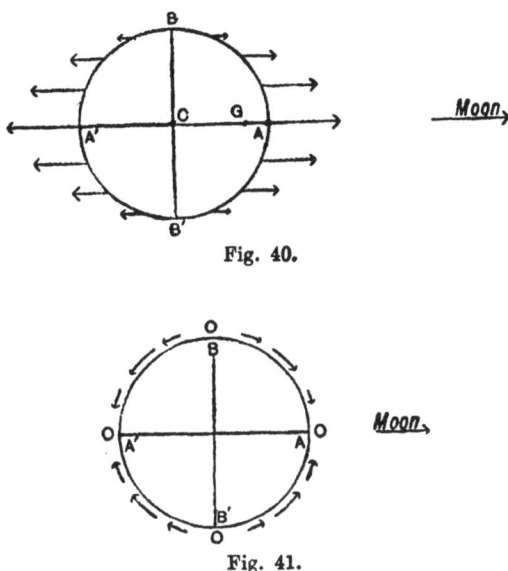

Fig. 40.

Fig. 41.

towards the Moon equal to *CG*, viz. that required to keep the surface matter in its circle, combined with a small extra pull in the opposite direction, and as the attraction at *A'* is less than that at *C* by very nearly as much as that at A is greater, the extra pulls away from the surface on the two sides at equal distances from AA' are very nearly equal, and we get extra forces over and above those needed for guidance in the circle, somewhat as represented in fig. 40.

Now to find the effect of these on the water in the canal we must resolve each along the vertical and horizontal. The vertical component merely diminishes the weight slightly and may be neglected. The horizontal is the important component and it will be seen that this horizontal component vanishes at B and B' and again at A and A', and is a maximum at the ends of diameters making about 45° with AA', somewhat as represented in fig. 41. These horizontal forces would move the water away from B and B' and heap it up at A and A', the ends of the diameter of the Earth pointing to the Moon, as represented with enormous exaggeration in fig. 42, *if the Earth always presented the same face to the Moon.* Though the Moon has no ocean the similar action of the Earth upon it undoubtedly deforms its solid body thus, and as it always presents the same face to the Earth it bulges slightly towards the Earth and bulges away from it on the opposite side.

Now consider the effect of increasing the rotation of the Earth from once in 27⅓ days to once in 24 hours. The Moon heaps up the water at two ends

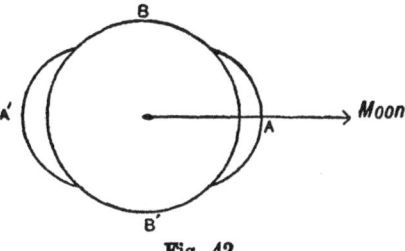

Fig. 42.

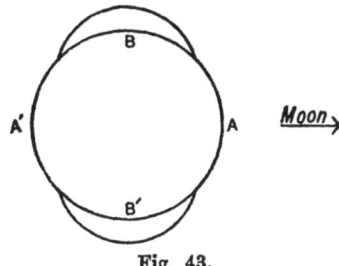

Fig. 43.

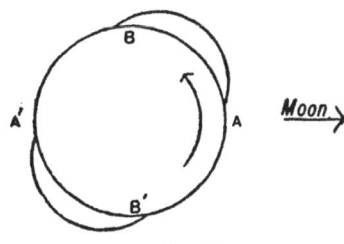

Fig. 44.

of a diameter, but the Earth is moving rapidly from west to east under the heaps. Relative to the surface of the Earth the heaps move from east to west. In the canal, then, we have two heaps and two hollows always travelling from east to west, two waves, each of length from crest to crest half the Earth's circumference. They move once round the Earth while a point on its surface travels from one position under the Moon to its next position under the Moon, i.e. in about 25 hours, and they bring to each point two high tides per journey round. As the circumference is 25,000 miles this implies a speed of about 1000 miles an hour.

But now comes in a curious consequence of the fact that the tides are waves. In a canal, or indeed in any sheet of water, waves once made and then allowed to travel on naturally, under the forces called into play merely by the shapes of the waves, have a definite speed of travel depending, if they are very long waves, on the depth of the canal. Each tidal wave is here 12,500 miles long, a very great length compared with the depth of the canal, which we have supposed three or four miles deep. Such a wave would require a canal 13 or 14 miles deep to have a speed of 1000 miles an hour under its own natural forces only. In a four-mile-deep canal the speed would be only about 550 miles per hour. The tidal wave, then, going round once in 25 hours and having a speed about 1000 miles per hour, is travelling much more rapidly than a natural wave. To get this greater speed the wave must so arrange itself on the canal that the Moon-forces shall conspire with the natural wave-forces to increase the speed of travel. The natural forces are always pressures from the crests towards the troughs of the waves, and if the crests of the tidal waves are at B and B' (fig. 43) and the troughs at A and A' instead of the reverse, it is seen from fig. 41 that the Moon-forces agree in direction with the natural forces, and so hurry the motion of every particle and increase the wave speed. The tide, then, tends to have its high water at B and B' as in fig. 43 and to be of such height that the two sets of forces give it just the right speed of 1000 miles per hour.

This tendency to have high water just $\frac{1}{4}$ way round the Earth from where we might at first expect it, is called the inversion of the tide. Were there no friction the inversion of tide in the canal we are imagining might be exact, i.e. low water might be directly under the Moon and on the opposite side and high water at right angles.

But friction acts in such a way that the Earth, turning counterclockwise as seen from the north, leaves high water and low water rather behind, rather to the west of the places where we might expect them, somewhat as in fig. 44, and we may

perhaps give some explanation of this lag as follows,
though the explanation is not quite accurate nor is
it complete.

In a water wave the water has a to and fro motion
as well as an up and down motion and about the
crest of the wave the forward motion is a maximum

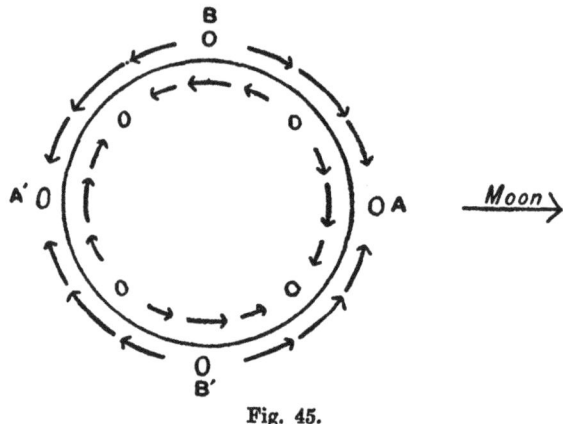

Fig. 45.

and it helps to transfer the crest to the next point
in advance. About the middle of the trough there
is the maximum backwards motion which helps to
transfer the hollow forward. The frictional resisting
force called into play by this horizontal motion is
a force on the water, backwards at the crest, and
forwards at the trough. We have to consider, then,

not only the Moon-forces, but also these frictional
forces, and the two sets together must conspire with
the natural forces to hurry the motion in the waves.
Remembering that the high water is somewhere near
B and *B'* (fig. 45), the horizontal friction-forces on
the water will be somewhat as indicated by the
arrows inside the circle, while the Moon-forces are
those outside; the friction-forces being much the

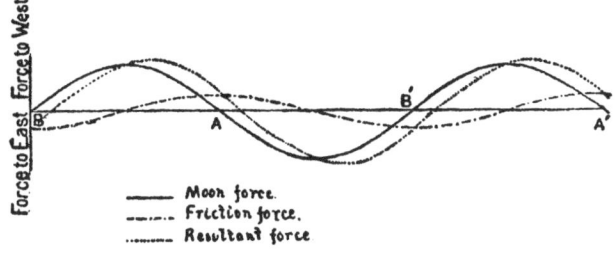

Fig. 46.

smaller set. The effect of their addition to the
Moon-forces is to carry the points of zero horizontal
force round from *BB'* towards the west, i.e. a little
way round in the clockwise direction and every part
of the force scheme will be similarly carried round.
This can be better seen, perhaps, if we represent
different points on the circumference by points on a
straight line, and the two sets of forces by curves
(fig. 46) where the curve representing the friction-
force should really be a little more to the right. In

P.   9

order that the compound of friction and Moon-forces may agree with the internal forces, vanishing at the same points and having maxima and minima at the same points, the waves must also be turned round a little in the clockwise direction, i.e. a little to the west.

The friction between the ocean and the Earth cannot of itself alter the spin of the Earth as a whole for the mutual action of two parts of a system cannot alter the sum total of their angular momenta. The friction acts indirectly by shifting the positions

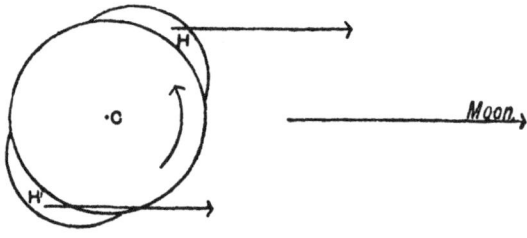

Fig. 47.

of high and low water, thereby altering the Moon's pulls and enabling them to put a brake on the Earth. This will be seen in a general way from fig. 47. The Moon-pull on the bulge at *H* is greater than that on the bulge at *H'*, since the latter is more distant ; the former tends to lessen the spin, the latter to increase it, and the net result is a diminution in spin.

To consider the effect a little more exactly, let

us continue the lines of the two forces at H and H' to the Moon's centre M (fig. 48). Only if the two forces were in the proportion of HM to $H'M$ would their resultant act through C the centre of the Earth, half-way between H and H'. But if the force along HM is represented by HM that along $H'M$ must be represented by a smaller length $H''M$, and the resultant is along $C'M$, where C' is half-way between H and H''. Now CC' is easily seen to be parallel

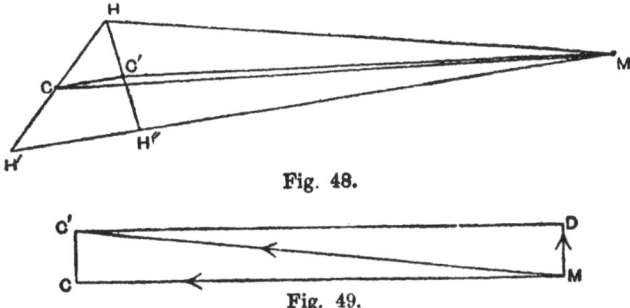

Fig. 48.

Fig. 49.

to $H'H''$ so that C' is necessarily above CM. Then the resultant action of the Moon on the two tidal heaps is a force not through the centre of the Earth, but on the H side of the centre. It therefore is what we have termed a sideway force, and it is always acting to slow down the rate of rotation. So we conclude that the tides are gradually reducing the spin of the Earth. After a time the Earth will move

so slowly that the tide will no longer be inverted
but it can be shown that it is then displaced by
friction in such a direction that the action still
reduces the Earth's spin.

Let us now turn to the Moon. In the first place
the action of the Earth in raising tides in the Moon
explains at once how she now turns always the
same face towards us, or rotates on her axis once
a month. When she was perhaps much hotter and
perhaps more plastic and certainly younger, the
Earth must have raised very considerable tides in
the solid body as well as in her oceans, if ever she
had oceans. On these the Earth would act as the
Moon acts now on the Earth tides, but much more
considerably. The resultant action would be a force
not through her centre, but a 'sideway' force op-
posing her spin round her axis; acting in fact as
a brake until the spin was reduced so far that brake
and wheel went round together, the Moon's period
of rotation coinciding with the month. The tides
on the Moon, tides in the slightly plastic body, are
always now at the same parts of her surface, directly
facing and directly opposite to the Earth.

In the second place there is a reaction of the
Earth's tides on the Moon equal and opposite to the
action of the Moon on the Earth's tides. We said
that this was a force on the Earth along $C'M$ (fig. 48),
so that the equal and opposite force on the Moon

is along MC' not quite directed to the Earth's centre. Resolving MC' (fig. 49) into MC and MD, the former being through the Earth's centre, the latter is a small component in the direction in which the Moon is moving in her orbit. This force is continually doing work on the Moon, tending to increase her velocity. But instead of this tendency being fulfilled there is an opposite effect. Inasmuch as without this pull along the path the Moon would be guided along a circle by the pull towards the Earth's centre; with the pull she moves slightly outside the circle, moves in fact in a slowly widening spiral, getting further and further from the Earth. As she gets further out, in increasing her distance against the pull to the Earth's centre she uses up not only all the energy put in by the pull along her path but also some of her own kinetic energy; somewhat as a cyclist going up a hill slackens speed, because the potential energy required is more than the energy which he puts in at the pedals, and so there is a call on and a diminution in his kinetic energy.

We conclude that in our idealised Earth with an equatorial canal, the action of the Moon on the tides is gradually lengthening the day, while the reaction of the tides on the Moon is gradually driving her out and lengthening the month.

On the real Earth, with its complicated distribution of oceans, the action is the same in kind but too complicated to allow calculation of the rate at which the action is going on. But there is a general principle which enables us to say what is the relation between the day and the month at any stage. This is the principle of the Conservation of Angular Momentum which asserts that in a system no interaction between the various parts will change its total spin. That spin is to be estimated by multiplying each pound of matter in the system by its distance from the axis round which the spin is to be calculated and by the component of the velocity perpendicular to the line drawn to the axis and adding up for the whole system. In the case of the Earth and Moon the spin is shared between the Earth and Moon. The Earth's share is gradually decreasing as the day lengthens. The Moon's share is gradually increasing, for her increasing distance more than makes up for her decreasing velocity, but the sum total for Earth and Moon is constant. Sir George Darwin has shown that the slowing down of the Earth's rotation will continue till the day is 55 of our present days. The month will then also be lengthened out to 55 of our present days and the Moon will be more than half as far again away from us as now. The Moon and the Earth will be always turning the same face towards each other, the tides will be at the same parts of the surface of each, and the tidal brake will cease to act.

So far we have left out of account the tides which the Sun raises. But these are by no means negligible. Every fortnight, when the Sun and Moon are in the same or in opposite parts of the sky, we have spring tides with high water much higher and low water much lower than at the times half-way between, when the Sun and Moon are at right-angles and we have the much smaller neap tides. The Moon tide is to the Sun tide about as 9 to 4, so that if the rise is represented by 13 when they are together, when they are opposite it is represented by about 5. The solar tides being so appreciable, the action of the Sun on these tides must also be appreciable and must tend to reduce the spin of the Earth. But the reduction is at a very much less rate than that due to the Moon. It will become more important in its effect when the Earth and Moon have come to an equal day and month which, as we have seen, works out at 55 of our days. For then the solar tides will slacken down the spin of the Earth still further without changing the length of the month and the Moon tides will again travel round the Earth. But now, relative to the surface of the Earth, they will travel in the opposite direction. The action between the Moon and tides will therefore be reversed, and the Moon will be gradually drawn inwards.

Let us now return from this vastly distant future

to the present day, and then travel back into the past. The process now going on implies that if we travelled back we should find the day and month both shorter and shorter and the Moon nearer and nearer to the Earth. And when the Moon was nearer the tides would be higher and the action greater. At last we should arrive at a time when, as calculation shows, both day and month were only from three to five of our present hours, and when the Moon must have been close in to the Earth. Here precise calculation ends.

If we make the very probable guess that before this the Moon and the Earth formed one body we can go one step further back in the history.

A planet of the joint mass of Earth and Moon, and of volume somewhat larger than the sum of the present volumes, as it probably would be, spinning round in about three hours would be very nearly unstable; the weight of the surface parts would only just suffice to hold them on to the surface. And we can assign a very probable reason for the small stability passing over into instability and disruption. If we could imagine a liquid globe to receive a deformation—to be pressed in, for instance, at the ends of a diameter and to be bulged out at the ends of a perpendicular diameter—and to be then released it would vibrate somewhat as a bell vibrates, and in a time depending on its density. That time of vibration works out for the liquid globe we have supposed at about 1½ hours. Let us suppose that for the actual globe it was something of this order, say it was two hours. The Sun would be raising tides in the globe, two tides in each day. And through friction these would be gradually lengthening out the day. A time might come when the solar tides, following each other at half-day intervals, would just agree in period with the period of free vibration. We should then have 'resonance' and the tides would become greater and greater until the crest of one of the waves—perhaps both crests, were thrown off to form ultimately a separate globe, the Moon.

There does not seem to be any wild conjecture in summing up the past and future history of the Earth and Moon system as follows. Immensely far back in the past there was a globe revolving round the Sun and spinning round with a day of very few of our hours. The Sun raised tides which gradually lengthened the day until their half-day period just coincided with the period of natural pulsation of the globe and the Sun-tides grew so high through this coincidence that the crests flew off and the Moon was born. At first day and month coincided, each being perhaps four of our hours. But the Moon raised tides and her action on these lengthened the day far more rapidly than the Sun could, while the reaction of the tides on the

Moon drove her ever further away. Meanwhile the Earth raised tides on the Moon which slowed down her spin until at length lunar day and month were the same, as they are now. In the past when Earth and Moon were nearer to each other the tidal action must have been more rapid than now, when it is so slow that even in 2500 years it is only doubtfully detected.

In the future it will be still slower. But we cannot doubt that it will continue till the Moon is half as far away again as now, till the month is twice as long as now, and the day is as long as the month, so that the Earth and Moon will present continually the same face to each other. But the history does not end there. The smaller solar action will continue to lengthen out the day without affecting the month directly and the lunar tides will then travel round in the opposite direction relative to the surface of the Earth. The reaction on the Moon will be reversed and she will gradually begin to retrace her spiral, this time towards the Earth, and perhaps at some enormously distant future time end her journey by reunion with the parent globe.

INDEX

For EU product safety concerns, contact us at Calle de José Abascal, 56–1°, 28003 Madrid, Spain or eugpsr@cambridge.org.

www.ingramcontent.com/pod-product-compliance
Ingram Content Group UK Ltd.
Pitfield, Milton Keynes, MK11 3LW, UK
UKHW010851090126
466816UK00011B/155